Elizabeth M. Almquist
CMU
June 1969

The
Science
Game

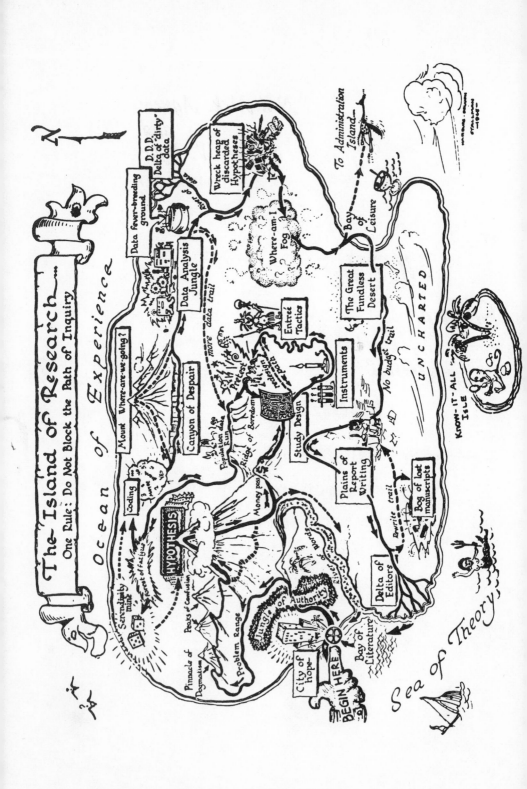

Neil Mck. Agnew
and
Sandra W. Pyke

York University

The
Science
Game

an introduction to
research in
the behavioral sciences

Prentice-Hall, Inc., Englewood Cliffs, New Jersey

PRENTICE-HALL SERIES IN EXPERIMENTAL PSYCHOLOGY

James J. Jenkins, Series Editor

DeBold, *Manual of Contemporary Experiments in Psychology*
Dixon & Horton, eds., *Verbal Behavior and General Behavior Theory*
Dustin, *How Psychologists Do Research: The Example of Anxiety*
Jakobovits & Miron, *Readings in the Psychology of Language*
Thompson & Schuster, *Behavioral Pharmacology*

© 1969 by PRENTICE-HALL, INC., Englewood Cliffs, New Jersey

13-795302-X

Library of Congress Catalog Card Number 73-76294

Printed in the United States of America

Current printing (last digit):

10 9 8 7 6 5 4 3 2 1

PRENTICE-HALL INTERNATIONAL, INC., *London*

PRENTICE-HALL OF AUSTRALIA, PTY. LTD., *Sydney*

PRENTICE-HALL OF CANADA, LTD., *Toronto*

PRENTICE-HALL OF INDIA PRIVATE LTD., *New Delhi*

PRENTICE-HALL OF JAPAN, INC., *Tokyo*

to our mothers who . . .

Preface

Some hold that the rules of the science game lie beyond the ken of ordinary men. Others maintain, quite as strongly, that "sciencing" is nothing more nor less than snooping around full-time. But this much is certain, that those who hold the first opinion by no means agree what the rules of sciencing are, and that those who maintain the second still treasure their white coats—the formal costume of the scientific sleuth.

We don't believe that the rules of sciencing lie beyond the ken of the nonscientist or embryonic researcher. Furthermore, with the growing impact of science on almost every activity, it is wise for all of us to become familiar with this great game—not only to appreciate its strengths, but also to understand its human limitations. For sciencing, like other games of consequence, is a mixture of art, enterprise, and invention held loosely together by man-made rules. The ingredients of any great game can be found in science—massive effort, goals, great plays, mediocre plays, lousy plays; umpires making clear calls, judgment calls, biased calls, and, of course, mistaken calls. There are prizes and penalties. You will find integrity, dignity, and deals, along with good luck and bad luck—but above all you will find commitment. So if you like strong games you'll like sciencing, as a player or as a spectator. However, if you like to play only on a smooth field, on a clear day, with cool umpires administering precise rules, then science is probably not your game.

It is presumptuous to pretend to know what impact, if any, a book will have. Nevertheless, it is not presumptuous to hope. It is our hope that nonscientists will develop a good feel for the science game and that students will develop an enduring taste for it. We hope that some of our colleagues will find material for discussion and argument in the pages to follow. Generally we believe that too often in our own attempts to teach research methods we have failed to place sciencing in its human and social context. The use of analogies—science as a game or as a news service—is an attempt to provide such a context. Also, continuing the attempt to place science in a social context, we have endeavored to provide an overview of man's many rules of evidence, from pub to pentagon, from laundry room to laboratory. In our view, rules of evidence are the rules and rituals, the methods and the magic we use to help us decide how, who, and what to believe, and for how long.

Want to play a game? It's called "How Do You Know?" You play it every day. When researchers play it, we call it The Science Game.

Someone has said there's a big gap between a bright idea and a finished manuscript. Even a little manuscript like this required many wise, patient, and kind people to help the authors over that gap. Some people will be surprised to learn they helped; some will be aghast; others will be relieved it's over. Each of the authors owes great debts to great teachers: to Professor A. Shephard, a warm evangelist for dust-bowl empiricism; to Professor H. Mehlberg, humane and scholarly spokesman for the universality of science; to Professor D. O. Hebb, who never let any mind sit down; to Professor Charles Osgood, the informal tutor on rhythm and range; and to Professor Norman Ward, the laughing wise man. We are also indebted to Professor B. Underwood and Professor D. Campbell, authors whose works we have taught from, learned from, and copied from.

Various people had the manuscript foisted upon them in part or in whole with special editorial burdens falling on Jean Black and John Cooke. For what must have seemed ad nauseam retyping of the manuscript, Mrs. A. Molaro and Mrs. L. Cantrell deserve special thanks.

We should like to express our gratitude to Professor James J. Jenkins, University of Minnesota, for his moral support and editorial help, to

Joan Brooks and Ed Lugenbeel of Prentice-Hall for their patience and for translating the manuscript into English.

It is said that aspiring authors are temperamental, distracted, and generally difficult to live with. We thank our respective mates for their tender, loving care.

Contents

The
Science
Game

one

Science
and Nonscience

MOST of the attempts to divide the world into two piles, like the good guys and the bad guys, or the left-wingers and the right-wingers, or the bald and the hairy, end up with two relatively little piles and a great big left-over pile for all the people who didn't fit into the two classes we started with.

The less we know, the easier it is to divide things into two piles. Such simple division usually requires that we play special tricks, or wear the blinders of bias or unfamiliarity that allow us to ignore an important similarity here and to magnify a minor difference there. Although dividing things into two piles is too simple and ignores differences, assigning each individual or event to a unique pile or category of his own is likewise too complex, ignores similarities, and leaves us with an unmanageable number of piles or categories.

In this book we will not end up with two separate piles, one for science and one for nonscience. All great human endeavors, like science, religion, art, and industry, have some general goals in common, such as the pursuit of excellence and the provision for growth and development. Nevertheless, science, religion, art, and industry differ in points of emphasis and means of expression. When we compare science to other undertakings, we will attempt to point out important similarities as well as note what we feel to be significant differences.

In this section, we will view science as a news service to see in what ways it is similar to and different from other news services. We will discuss, too, how some of the misunderstandings between science, art, and religion arise over words. We will also look at the charm and power science has for its addicts. Finally, we will refamiliarize ourselves with some of the major rules of evidence we employ every day in pub and pentagon, in laundry room and laboratory.

chapter 1

This Thing Called Science

What is science? One way to answer the question is to look at what a cross section of scientists do, or to read everything that has been written about science. Such an approach would take too long. We could look at a brief summary, but a summary is often quite meaningless except for the person who worked so hard to condense all the material. Here is another alternative. Assume that we have access to a great metal drum full of slips of paper containing statements about what scientists do and descriptions of what people have said about science. Now let us mix the slips well, reach in, and draw a sample of statements from different parts of the drum. Although this approach will not tell us as much as watching all scientists and reading everything that has been written about science, it should, nevertheless, be more interesting than just reading a summary. We may even come up with a few surprises!

A SAMPLE OF SCIENCE

What does the sample of slips we have drawn out of the drum have to say? "Scientists agree on new finding." Nothing surprising about that since they are supposed to be able to agree. "Scientists find temperature on Venus too hot to support life as we know it." Pretty remarkable—measuring the temperature of Venus. "Scientists still search for cancer cure." Apparently science cannot guarantee fast results. "Scientists disagree." Scientific truth, then, does not always come clearly labeled; what is true and what is not true is not always agreed upon even by scientists. "Scientists say problem unsolvable by methods of science." So the reach of science is somehow limited. "Scientific literature grows at fantastic rate!" A scientist cannot keep up with the literature in his own field. "Doctors ridicule and banish colleague who later is proven right." So the truth is not always welcome, or recognized as such even by those claiming to be scientists.

The Human Side of Science

Our small sample of statements gives the impression that science is a very human enterprise—human in the sense that it has its limits, that even scientists disagree, that even scientists resist new ideas and new findings. Apparently there is more to science than mental giants combining with magnificent electronic brains to produce neat packages of 100 percent pure truth. Much of what has been written about science has presented her dressed in her Sunday best. Since this tidied-up view of science has been more than adequately represented, we will approach science as she appears during the rest of the week: her Monday morning blues and her Saturday night binges.

The scientist, like the rest of us, faces the problem of developing enough faith to make the next project worth doing, but not so much faith as to blind him to all news, or evidence, other than that which he wants to hear. In order to reduce the biasing effect of his blinders, he usually follows certain written and unwritten rules. Nevertheless, he always runs a variety of risks, and develops various tricks of his trade as a consequence. In our view, there is no single agreed-upon "scientific method." Instead, there are mixes of rules, strategies, tricks, and tactics upon which the researcher draws. The particular mix selected for a given project depends upon the personal preference of the researcher and upon the specific gremlins and risks he faces in the particular study —different risks require different handling.

SCIENCE AS A NEWS SERVICE[1]

A useful analogy can be drawn between the activities that are involved in producing a newspaper and those involved in producing a scientific report. First we will examine a news story, then we will look for parallels between the problems facing the reporter and those facing a researcher.

The News Reporter

A newspaper reporter sent to cover a sports event cannot report everything that happens because he does not see everything that happens. Neither can he report everything he sees because it would require too long to write it up and it would take too much space. Therefore, he selects certain things from all he observes, and weaves these into a report or story. He also emphasizes certain parts of the story and plays down other parts. And the title of his story highlights selected points.

The sports event he covers may be a key football game between the Giants and the Packers. For three quarters it has been a seesaw battle between the Giants' great quarterback and the Packers' hard-running backs; at the beginning of the fourth quarter the Packers are ahead by three points. Early in that quarter the Giants' regular quarterback is hurt and is replaced by a newly-acquired second-string quarterback, a great college star without much professional experience. The first time the new quarterback gets the ball, a Packers' linebacker breaks through, smashing him for a twelve-yard loss. The crowd starts heckling, and then the new quarterback's first attempted pass is intercepted.

One Man's View. At the end of the game the Packers have won by three points and the reporter turns in his story. His report emphasizes the failure of the Giants to gain points in the fourth quarter, and the story headline reads "New Quarterback Fails under Fire." Now, many good first-string quarterbacks have been thrown for big losses, have been heckled, and have failed to gain points in fourth quarter play without being blamed for loss of the game. Our reporter, however, decided that the Giants did not get points in the fourth quarter because the new quarterback could not stand up to being thrown for a loss and to the heckling from the crowd.

Other Views. There are alternative interpretations, some ob-

[1] Henryk Mehlberg, *The Reach of Science* (Toronto: University of Toronto Press, 1958), pp. 12–15.

vious and others not so obvious. One could be that had the Giants' regular quarterback been in the game, the team still would have done no better. Another interpretation could be that the new quarterback would have done better had he not fought with his wife just prior to the game. (His wife wanted him to quit football and go into business with her father rather than risking injury and being away from home.) So, perhaps it was not just the yardage loss and the heckling which accounted for his poor performance. These things were mixed with the frustration and anger growing out of the argument, and the combination of events helped send the pass over the receiver and into the arms of one of the defending backs.

Still another interpretation appeared in a different paper. The reporter for this paper concluded that the Packers' clever coach had finally worked out a strategy for getting his linebackers and defensive line to locate chinks in the Giants' huge front wall.

The important point is that there are always several plausible explanations to consider for any event, whether you are a reporter or a researcher.

Science and Its Reporters

Science, like other news agencies, has its researchers who collect and report news. Like newspaper reporters, they must make decisions about what aspects of their experiment or research should be reported and which of these should be emphasized. They too have problems in deciding which of a number of possible conclusions is most warranted.

The researcher, at times, has advantages over the newspaper reporter in that he has certain techniques which allow him to rule out some of the alternative interpretations. Such techniques are equivalent to giving the reporter the magic power to have the game replayed under different conditions, for example, without the wife having upset the quarterback. Nevertheless, many alternatives usually still remain to confound him, even after these techniques have been used.

The Editorial Writer

Not only does the press have its reporters, but it also has its editorial writers whose job it is to take an overall view or to see what patterns emerge from a series of the stories turned in by reporters. Thus the sports editor may decide, on the basis of a series of games, that the new quarterback is good. He thinks the real problem is that the Giants need to replace some of their perennial front-wall stars with younger men. The coach and front-office staff have been negligent in not bringing younger men along. The editor relies on the observations turned in by several reporters. By pooling them he reaches his conclusions.

Science and Its Editors

Like the reporter, the researcher must also face editorial policy. The news analysts and editors for science are the theorists, and the journal editors and their consultants. These analysts critically evaluate the individual researcher's reports to see how they fit in with more or less established observations or theories. If a scientist's observations and interpretations are way out, he may have difficulty getting into print. As the chemist Conant pointed out: a few facts that do not happen to correspond with current theory do not result in the immediate overthrow of the theory.[2] Just as the news analyst and the editorial writer are usually more influential than the fact-finding reporter, so, in science, the theorist is usually more influential than the individual researcher reporting relatively isolated facts. While Einstein, the theorist, is known to every school child, the names of Michelson and Morley, on whose work Einstein relied heavily, are relatively unknown.[3]

THE TRADEMARKS OF SCIENCE

Although we have been stressing the similarities between science and other news agencies, it should be noted that there are differences as well. What are the main trademarks of science?

Just the Facts, Please!

One of the most obvious characteristics that sets science off from most other news agencies is its great emphasis on the accurate reporting of what the researcher observed. A newspaper reporter must maintain a certain level of accuracy in his reporting, but a scientist, if he follows the rule book, must be scrupulously exact in his reporting. A scientist cannot report everything that happens because of his inability to either observe it or record it. He nevertheless gives special attention to being accurate in what he does report. Thus, as a scientist, you will not report everything that happens, and you may not be right in your conclusions as to why something happened, but you are expected to report your observations very carefully, and in a way that

[2] James B. Conant, *Modern Science and Modern Man* (Garden City, N.Y.: Doubleday & Company, Inc., 1953).
[3] Mehlberg, *The Reach of Science*, p. 13.

is not open to argument. It is assumed that had another scientist been observing the same event, he would verify your observations even though he might not agree with your conclusions. The ideal is to report your observations so objectively that even a mother-in-law would have to agree! Observations about which different observers or measurers agree are called *reliable observations*. When researchers ask about interobserver or interjudge reliability, they are asking whether different researchers or reporters agree about a given observation.

Returning to the analogy of the events at the football game, we presume that different observers would agree that the quarterback was thrown for the loss, that his pass was intercepted, and that the Packers won the game by three points. However, the same observers might well differ in their explanations of why the Packers won the game. Thus scientists may not fight over their observations (e.g., that the patient got better) but they will fight over their explanations (e.g., that it was the new drug, or that it was the bed rest).

If newspaper reporters went to the time and trouble that researchers do to ensure that their observations are reliable, our news would be more than one day old, and the newspaper would have to double its space so as to have at least two independent observers covering each story. Those aspects of the story about which the observers disagree would not be seen as reliable news, but rather would be seen as commentary, opinion, or interpretation. Therefore, in order to encourage objective reporting, scientists are allowed more time and space than are newspaper reporters.

I Wonder What Will Happen if . . . ?

There is another difference between newspaper reporters and scientists. Scientists like to experiment. Reporters presumably do not actively manipulate the events they observe, but researchers feel happier working with events that they can manipulate. This controlling of events permits researchers to observe the different outcomes when they alter the action of key players. The trademark of science today is being able to experiment with events in addition to, or instead of, merely making observations. Nowadays learning the art and skill of experimenting usually takes eight to ten years of university training and indoctrination.

Worship That Generalization

Still a further difference exists between science and a news service. Scientists are not nearly as interested as the newspaper reporter in the immediate present.

As the philosopher Mehlberg points out, probably the major difference between science and other news agencies is the scope of scientific information and the many uses to which it can be put. "The ordinary news-agencies are concerned exclusively with single, particular, contemporary, and local facts, with the 'here and now,'"[4] whereas scientists attempt to produce information which is of lasting value and which can be generalized—information which is widely applicable. Consider the statement, "Metal expands when heated." This is not a here and now fact alone; it is a piece of information that applies equally well in Timbuktu and Salt Lake City. Presumably metal will expand when heated next week, next year, and in the next century. It is a piece of information that can be generalized widely through time and through space, beyond the here and now. Let us assume for a moment that different pieces of iron were as temperamental and unpredictable as some movie stars. No longer would we be able to make the generalization that iron expands when heated. Rather we would be faced with a series of particular statements such as: "Yesterday Professor Jones reported that a piece of iron expanded when placed in a hot oven for a half hour," and "Today, Professor Williams reported that his piece of iron contracted when put in a hot oven for a half hour." Here we have information or news of a very local nature—local in the sense that it will be very difficult for us to predict what will happen if somebody else puts a piece of iron in his oven. The goal of science is the search for durable information that can be broadly applied in time and space beyond the here and the now; other news agencies are content, in the main, to report news of a more local and time-limited kind. The phrase "as out of date as yesterday's newspaper," would be one of the cruelest things you could say about a scientist's work—particularly about his observations; these, presumably, should withstand the erosion of time even though his theories may not.

Thou Shalt Not Fool Thyself Too Much

There is another point of emphasis in science that differentiates it from most other endeavors: the elaborate methods scientists employ for keeping to a minimum the number of ways they have of fooling themselves about their scientific conclusions. A scientist attempts to be his own impartial judge, prosecution, defense, and jury. It is an ideal continually sought, but seldom attained—but more of this later!

The Language of Science

There is still another difference between science and other news agencies. Scientists, unlike other newsmen, are usually extremely

[4] Mehlberg, *The Reach of Science*, p. 14.

fussy about the language they use in describing their work because other scientists often attempt to repeat the work to see if they get the same results. If the directions are poor they become disgusted. The urge to kill arises in the most pacifistic of us when, after deciding to build a model satellite for a seven-year-old, we find that many parts have not been labelled clearly and that the instructions merely state, "Now glue together the type B stabilizer and the type C stabilizer, attaching both to the underside of the solar battery." We are frustrated by indecision, without a clue as to the shape or the size of any of the three items. It is like being told to go to Grand Central Station to find Harry and take him to meet George, and then take both of them to meet Mike, when you have never seen any of the three men before.

Scientists abhor any vagueness in their recipes. Their aim is to write the recipe in such detail that the person following it has to make few, if any, decisions. This scientific ritual tests whether observations are limited in time or space; other researchers in other places and at later times, following your procedures, should be able to make the same observations that you did.

Words and Ghosts. While the news reporter is expected to stick to the facts, the editorial writer deals in speculation and abstractions. So in science the front-line researcher is expected to stick to what he has seen, while the theorist deals in speculation and generalization. Many scientists feel ill at ease in having to deal with abstractions, with words that cannot be tied to events or objects that can be clearly seen. For example, before the discovery of the microscope made germs visible and therefore scientifically respectable, some men believed that there might be small, invisible somethings that made people sick. Until the microscope produced visible evidence, such ideas remained vague, and could only be tied to little ghosts. Similarly, the notion of tiny particles of matter moving rapidly through space became respectable only when it was supported by events observable on photographic film, or by a tracing in a cloud chamber. Scientific theories include many ghost-like terms which are a source of fascination for some scientists (the theorists), and a source of embarrassment and annoyance to the other scientists (the researchers). Until the events to which the ghost terms refer can be photographed, weighed, or transformed into visible form, they remain suspect. When they become visible, the ghost terms have been transformed into operational definitions.

Therefore, we can see that while science places great emphasis on reality and visible events, on concrete or operational definitions, it must deal in many instances with a world of make-believe. But the eventual aim of the scientist is to find some suitable tangible events to which he may tie his abstraction. The methods of blending the concrete world with the world of make-believe is the most fascinating part of science.

So far, we have noted certain similarities and differences between science and other news services. We have stressed the emphasis science places on careful reporting, experimentation, findings that hold beyond the here and now, elaborate safeguards against fooling itself, and attempts to tie all its words to observable events.

Other Rules of Language Usage. There are, of course, many rules governing the use of words and language other than the rules used by the researcher, which are personally and socially relevant. For example, to express personal feelings we may say, "I feel sad." It is extremely difficult to agree upon a definition of sadness, a fact aptly demonstrated by a popular song, "Laughing on the Outside, Crying on the Inside." The proverbial tears of joy also help to confuse the picture. Yet here we have an instance of a word like "joy," which has great personal validity and significance but which, like other mood terms, has resisted being tied to visible or observable events that always apply to a wide variety of people. Thus, in addition to research rules, there are more or less private rules of language usage which you follow in talking to yourself—rules like those followed by Humpty Dumpty, whose words meant whatever he wanted them to.[5] Similarly, you and a very close friend may develop a more or less private language which the two of you accept as meaningful, but which is meaningless to other people. Therefore, we see that the personal and semi-private use of language often involves different rules than the scientific use of language.

Similarly, in some instances the spiritual and religious rules of language usage differ from those of science. Terms like "God," "Heaven," and "Hell," significant though they may be on the personal and social levels, have no broadly accepted, tangible reference points. They lie beyond the reach of science because it is difficult even to conceive of any microscope or other aid that would make them visible in the same sense that germs and planets are.

Where is the Supreme Court of Language? There are some scientists who feel that everyone should adopt the rules of the science game as far as the use of language is concerned. And there are others who feel that no one group has any special right to set themselves up as the Supreme Court of Language. Thus, rather than thinking in terms of one set of rules being superior to another, some people accept the idea that there are different rules for the use of language—scientific rules, personal rules, and religious rules. These various sets of rules exist for different purposes and are therefore legitimate within their respective areas of discourse.

[5] Lewis Carroll, *Alice's Adventures in Wonderland and Through the Looking Glass* (London: Wm. Collins Sons and Co. Ltd., 1939), p. 187.

SCIENCE AND NONSCIENCE

We have provided no clearcut way of separating science from nonscience, but we have suggested points of emphasis. To the extent that an investigator deals with events that can be pointed at, uses language objectively, and makes observations that extend beyond the here and now, he approaches the goals of science. To the extent that an investigator deals with events that cannot be pointed at, uses vague language to describe such events, and concerns himself mainly with observations of the here and now, he moves away from the goals of science. When a researcher starts toying with a new idea and making his first preliminary tests, he operates in a vague, here and now world. His goal, however, is to move more and more toward providing objective information that can be of general use.

Another way of defining science would be to say that it consists of facts fitted into a framework which permits the prediction of more facts and so helps to create a more general framework—which leads to a still greater number of facts—and so on and on. Thus facts heaped on a table do not constitute a science. We have a science, however, when facts are assembled in a way that: 1) matches what we think we know of the world, and 2) leads us to discovery or invention. Some scientists, the researchers, concentrate more of their time on providing the facts; others, the theorists, on building the frameworks into which the facts may be fitted.

THE SCIENTIFIC PASSION

But what are the lures and charms of science? What keeps scientists at it in spite of hard work and wrong guesses?

Creating a Self-Satisfying Picture

Scientists do a lot of imagining. They create images, pictures and stories about the past. There is astonishing gratification in creating a new and personally satisfying story about some past event, whether it is an important historical figure, a prehistoric animal, the location of Troy, the laws of heredity, or even why the strange lovely girl smiled at you. Certainly, creating a self-satisfying picture of the past is one of the lures of science.

Similarly, a rare excitement comes from being able to create a personally satisfying picture of the future. The passion of the ages has been to attempt to predict the future—little parts of it, big parts of it. This is the lure of the experimental sciences, to attempt to predict the existence of a great source of energy tied up in tiny particles, to predict the temperature of Venus, to predict which of the old illnesses a new drug will cure, to predict marital adjustment or job success. Armchair forecasting can be a most satisfying occupation.

But is there no more to science than creating personally satisfying stories of the past and the future? Yes, but never underestimate the critical importance of these stories in directing the course of science.

Creating a Scientifically Plausible Theory

The next great source of gratification occurs when fellow scientists become excited about your particular view of certain past or future events. In science, however, as in other fields of endeavor, the excitement that greets a new idea is not necessarily all positive; the rewards range from the Nobel prize to ridicule and expulsion.

The fights and intrigues between individuals, and also between great institutions, that grow out of attempts to introduce new views of the past and future, make fascinating reading. How ludicrous it would be if the training of young scientists included a course on "scientific public relations: how to smoothly introduce a new idea in science." Ludicrous? Perhaps.

Let us consider the case of Ignaz Semmelweis, a Hungarian physician who, in 1847, created a new picture of why so many women died with fever following child birth.[6] His view was that the fever was caused by his colleagues carrying something on their hands from the post mortem room to the new mothers. He even demonstrated there was a decrease in mortality from the fever if the doctors washed their hands before examining patients. Nevertheless, his medical colleagues refused to believe him; he was fired and ridiculed for his efforts. Finally he died in an insane asylum, thus conveniently suggesting to his enemies that he had been crazy all along.

William Harvey also suffered ridicule when he introduced his theory that the blood was actually circulating in the body, rather than ebbing and flowing as was then the generally accepted idea.[7] After twenty years, and a lot of abuse, the new view was generally admitted.

Even though Harvey was ridiculed, he was a better strategist than Ignaz Semmelweis. Harvey dedicated his book to King Charles, as

[6] W. I. B. Beveridge, *The Art of Scientific Investigation* (New York: W. W. Norton & Company, Inc., 1957), pp. 113–14.
[7] Beveridge, *The Art of Scientific Investigation*, p. 106.

Beveridge has pointed out.[8] The tactful doctor drew an analogy between the heart and the body, and the monarch and his realm. Poor Ignaz, unfortunately, did not choose a king to be on his team.

But is there then no more to science than creating a picture of the past or future that other scientists agree is possibly correct, and so worth working on? Yes, but never underestimate the importance of the opinions and biases of the public and of other scientists in directing the course of science.

The Acid Test

We have discussed the gratification of creating a new picture of the past and future, a picture that is personally satisfying and is also considered plausible by other scientists. But the ultimate scientific satisfaction lies in something else, in the rare thrill of having a particular picture or theory show a surprisingly good fit with the facts as they are produced by the proponent and other scientists.

Oh, what delight for the scientific soul to have an archeologist find Troy where he said it should be! What a complete sense of fulfilment to have successfully predicted that there was enough power inside a piece of rock no bigger than a fist to light a modern city. What self-justification to have foretold that smallpox could have been prevented by a wee taste of the bug itself.

To predict the unexpected, and to have other scientists provide evidence so compelling that your competitors agree you were right—now there's a path to glory.

Evidence: Hard, Soft, Softer. But this is heady wine without a prior feast of basic vittles. For example, you may be surprised to learn that scientists do not always agree about what is and what is not compelling evidence. Quality of evidence changes with time. It may be useful then to think of observations about which scientific competitors can agree as "hard" evidence. Observations upon which only friends agree can be thought of as "softer" evidence. For example, upon digging down where you, the scientist, said Troy would be located, you may find the ruins of three medium-sized buildings, pottery, and tools typical of the period. Thus you and your friends have found evidence supporting your picture of the past. Your rivals (people with a different picture of where Troy should be) reply, "You call the remains of three little shacks evidence of Troy? By luck you hit some sheep herders' huts!" You answer that you have probably found the edge of Troy and with additional digging you will find the major part of the city. However, there

[8] Beveridge, *The Art of Scientific Investigation,* p. 111.

is no more money immediately available for further excavation. So one group, you and your friends, wait on the edge of Troy for more money. The other group, your theory rivals, see you cherishing old sheep herders' huts and waiting to dig down 30 feet at terrific expense only to find antique sheep dung. For those involved, the mystery is compelling. Each new bit of research, each new suspect and new intrigue contributes to the unfolding plot, and lures you on and on and on.

chapter 2

The
Currency
of Science

Science can be thought of as a set of recipes or statements about the world in which we live. The main currency of science is language and yet certainly not all language is scientific. We noted in Chapter One that there were different types of language, or language which followed different rules; some language follows personal rules, some language follows religious rules, some follows mathematical rules and some follows scientific rules. To get a clearer picture of the inner workings of science and research it is necessary to take a closer look at language.

The study of language as it is traditionally taught can be dry, academic stuff. Yet there is probably nothing more practical than a working knowledge of the different levels of language in the world around you. Understanding the power of language is very practical, whether you are a teenager on your first date, a politician trying to round up the uncommitted vote, an advertising account executive, a student writing a final exam, a job applicant, a committee man, a panel participant, or a scientist presenting a new finding or theory.

THE MANY TONGUES OF SCIENCE

What is language? Stripped to its bare essentials, language is nothing more than a series of marks on paper, or a string of sounds, that follows certain rules. One of the rules is that you be able to tell one mark from another and one sound from another because usually these marks and sounds stand for different objects and events in the world around us. We don't worry about the specific qualities of the mark selected to represent a given object; all that is necessary is that we know which mark represents or is tied to each object.

Marked Balloons

One way of thinking about language is to think of words or marks painted on balloons floating above the real world. Each marked balloon is supposed to represent some object or event, and should be tied to the object or event it represents. Some marked balloons are tied to or represent only one specific object.

Others represent whole classes or groups of objects or events.

Learning a language involves at least two things. It involves learning to produce the marked balloons (written or spoken), and it involves being able to tie them to the same objects and events as do other people who speak the language.

With many marked balloons there are no problems in knowing what objects and events they represent—we all tie them to the same objects. Such is the case with clocks, cows, inches, jellybeans, cars, and kippers. Other marked balloons, however, present some difficulties and we won't agree about where they should be anchored. Examples of words about which people differ include:

In the case of such words, we no longer tie them to objects by following the simple group rules—we follow personal rules. When you say, "I saw a 1964 black Ford Falcon," anyone listening has a good idea of what you were talking about, and to prove it could point out the type of car you saw. (Pointing is analogous to tracing the string from the word to the physical object car.) However, if you said, "I saw the most beautiful girl in the world," we would have trouble pointing her out without help from you. Eventually we might learn that most girls you tied the "beauty" balloon to were tall redheads. We are learning some of your personal rules for tying language to the objective world. Thus, not only are there common rules covering the use of some words (e.g., "cows" and "cars"), but there are also personal rules covering the use of other words (e.g., "beautiful" and "ugly").

In some cases people can figure out your personal rules. For example, by having you point out a series of beautiful women, they might find that your rule for using the word "beautiful" is to tie the word to tall redheads. In other cases, however, it is not so easy to detect what rules an individual is using—when you can't point to the object or event which the word represents, when you can't see the object or event to which the marked balloon is tied.

Dream Balloons

If you have ever attempted to describe a fantastic dream to anyone you are aware how much was lost in the telling. Why? Because the words are tied to images and feelings inside you that can't be taken out and pointed at to show all their grotesque detail and intensity.

You have several courses of action in your attempts to communicate your dream to others. If you are like most of us you will say something similar to "Boy, did I have a terrible dream last night!" and your friends will reply with mild grunts to acknowledge that they hear you. You wish you could show the dream in cinemascope with full sound so that your indifferent sidekicks would give forth with more than mild grunts of indifference. You can attempt to elaborate your dream by painting a word picture for them. You can string together some marked balloons about "purple snakes the consistency of raw liver," and "sharks with corkscrews sticking out of their mouths," and "other ugly things you can't describe." Now your friends begin to show a bit of interest. You're not sure how effective you've been in communicating with them because the strings from your marked balloons run into their heads and disappear— one end of the string tied to your words, the other end tied to their own personal images that can't be simply taken out and pointed at or lined up and compared with your images. If we could give the strings a yank and have their images pop out onto the table then it would be easy to correct some of their misconceptions about your dream. You could then point out that the purple snakes with the consistency of raw liver were not at all like the image we pulled out of their heads. Why the images of the snakes produced by one of your friends are the size of garter snakes and have quite friendly brown eyes and sad mouths. However, the snakes in your dream were the size of dinosaurs, but without arms and legs. The green eyes (all eighty of them) were mounted on long scaly tentacles and the mouths leered and dripped blood. If we could compare images directly like this, you could then tell your friend to take his scrawny image back inside and come up with a real snake.

There are many other instances in which words are tied to personal images and feelings which can't be pointed at—images of heaven and hell, feelings of fear and sadness, of anxiety and peace.

Thus, when the objects (penguins or paper clips) or events (9 A.M. or graduation) to which words are tied can be pointed at and compared, we can tell whether we are both using words to represent the same things. However, if we can't point at the objects (ghosts and gremlins), and events (crying on the inside), in question, we have no way of telling whether we have all tied the word to the same object or event. For example, if we all agree that we are "sad," we assume that we mean the same thing—that the word "sad" is tied to the same image and feeling inside each of us. If we could pull the string to lay all the "sads" on the table, we'd expect to see similar things. We'd probably be quite surprised if one "sad" turned out to be overactivity of a nerve in the throat, plus an image of eight long violin strings playing ballads with the aid of the prairie wind, whereas another "sad" was a tension in the chest muscles, and an image of an ocean crying against the shore.

The statement "I feel sick" is understandable when the patient is standing there bleeding profusely from a large wound. The "sickness" can be pointed at. However, when no such obvious sign exists, the doctor has the job of tracing the string from the word "hurt" into the patient. He has the job of reaching into the patient and pulling out some visible sign such as reaching in and bringing out the patient's temperature, or reaching in and bringing out his blood count, or (if he can't find anything else) he may reach in and bring out his appendix.

Summarizing then, we see that in some cases language is tied to objects and events that can be pointed at—the objective world. Groups of people agree that certain words should be used for certain objects and events. In addition, groups and individuals are allowed some freedom in applying other words to parts of the objective world; one of Damon Runyan's characters says, "Let's pick up a couple of flowers [girls] and hoist a few smiles [drinks]."

In addition to the objective world, language is tied to subjective images and feelings which can't be pointed at and compared directly. In such cases we often assume that the same words have the same meaning for each of us. In other cases we try to bring some semblance of the inside images to the outside in the form of tears or smiles, of music, drawing or painting, dancing, singing, sighing, or talking. We don't accept these as foolproof signs of what's going on inside, as the following quotations show, "sour grapes," "still waters run deep," "you never really get to know him."

In examining science and research, it is particularly important to keep in mind the different levels of language. It has been proposed that language consists of sets of rules for combining signs or symbols (written or spoken). We observed earlier that there is no one supreme court deciding what the rules are. Some rules are set by textbook writers, some by curriculum makers, some by groups of editors, some by you. Most language, however, can be considered as belonging to one of three broad types: pragmatic (subjective or personal), semantic (objective or tangible), and syntactic (structural or grammatical). It is important to take a closer look at these three types of language if we are to appreciate some of the artistry as well as some of the great battles in science.

One thing should be made clear. There is nothing magical about these rules; they are man-made in the same way that rules for playing football, chess, or tiddlywinks are. They decided to call this event (a certain mark on a photographic plate) an "electron." They could have used any one of a million labels such as "flurd." However, the label "electron" was selected, and if you're going to play the physics game you must learn what label goes with what event so that you can use the label when you want to talk about that event. Similarly, in the syntactic use of language in mathematics and logic, men decided the rules, and then accepted them; they decided that this mark plus this mark will equal that mark. Mathematics is a game for combining marks in different ways. If you're willing to play by the rules, you can play the game, otherwise you are penalized or perhaps expelled from the usual playing fields or journals. In that case, you may go off and set up your own playing field and start your own game or journal.

THREE TYPES OF LANGUAGE

Personal Rules (Pragmatic)

"When *I* use a word," Humpty Dumpty said in a rather scornful tone, "It means just what I choose it to mean—neither more nor less."

"The question is," said Alice, "whether you *can* make words mean so many different things."

"The question is," said Humpty Dumpty, "which is to be Master—that's all."[1]

Humpty Dumpty is right. Language in some cases follows the rules or inclinations of individuals and can lead to wonderful nonsense.

'Twas brillig, and the slithy toves
    Did gyre and gimble in the wabe:
All mimsy were the borogoves,
    And the mome raths outgrabe.[2]

Children come up with delightful expressions: "that hill is too big for my shoes." Love mumblings are another case in point: "Isapeachy-

---

[1] Carroll, *Alice's Adventures in Wonderland*, p. 187.
[2] Carroll, *Alice's Adventures in Wonderland*, p. 189.

cabbageluvacurlyturnipman." Poets too subscribe to personal rules in their use of language:

> Go and catch a falling star,
>   Get with child a mandrake root,
> Tell me where all past years are,
>   Or who cleft the devil's foot . . .[3]

The foregoing all represent one very important level of language usage where the rules are personal and flexible, where the words can be vague or abstract, and unique combinations are acceptable. Personal rules of language usage are wonderful for poetry, lovemaking, and play. However, when it is necessary for large numbers of people to coordinate their activities, more standardized rules of language usage are required. For the purpose of research, personal rules are usually considered unsuitable. If you want to play the research game, you have to learn research rules which involve objective (semantic) rules and also mathematical and logical (syntactic) rules.

Objective (Semantic) Levels

Whereas the pragmatic level of language just considered relies heavily on personal rules, the semantic level operates upon shareable rules for tying words or labels to clearly seen objects or events. This level of language deals with obviously solid things like cups and clocks, cars and cows—with objects that can be pointed at, that can be seen and agreed upon even by members of different political parties, different sexes, or different religions. The semantic level of language deals with the empirical or shareable parts of our world, those which are open to public inspection.

The Democrat and the Republican can readily agree that the candidate spoke for 30 minutes. That is a semantic statement. However, they may never agree whether it was too long or too short a speech, or whether it was a good or a bad speech. This usage of language is at the pragmatic or personal level, the level of group loyalty. The semantic rules are designed for people with different personal values and group loyalties. These rules are extremely important in research and science; however, these rules limit your vocabulary. Personally meaningful words like "beautiful," "ugly," "absolutely right," "irrevocably wrong," "heaven" and "hell," "love," "sadness," "hope" and "glory," "liberty," "Schmaltz," and "cool, man," do not at present belong at the semantic level.

While researching, you try to play by semantic rules of language. Of

---

[3] John Donne, "Song," in *Patterns for Living* (Part I), eds. O. J. Campbell, Justine Van Gundy, and Caroline Shrodes (New York: The Macmillan Company, 1940), p. 227.

course, researchers and scientists still use pragmatic rules as well. If they did not shift back and forth, they would be intolerable to live with and work with. Imagine living with someone who played only by semantic rules. The dialogue might run something like this:

You: "You know I'm really beat."
They: "You mean your blood sugar level has fallen below level six."
You: "Maybe that's what's happened. Say, look at that terrific sunset! It's breathtaking."
They: "In terms of color spectrum it includes wave lengths from 5600 to 6100 angstrom units. However, I don't detect any great changes in my respiration rate, perhaps four cycles per minute above the average for this time of day."
You: "Let's turn the verbal decibels below the audible range before you make me sick."

Just as the semantic level of language can be out of place and inadequate in the pragmatic playing fields—the pub, the theater, the cocktail party, the bedroom, the political convention—so the pragmatic level of language can be most inadequate in the playing fields of science and research. Imagine attempting to do research with someone who played only by the pragmatic rules:

You: "How did you create such a mammoth mushroom-shaped explosion?"
They: "It wasn't easy, Dad—a dash of some black, shiny stuff—a pinch of some weighty water, bring them just out of tickling distance and don't jiggle my elbow, unless you really want the world to blow its roof off."

There might, of course, be an advantage in wording our scientific findings in this way: at least no one could ever steal this secret. We all know of the woman with a special recipe for chocolate cake, or a man with a unique recipe for a dry martini. They share it with you gladly in a language designed to be vague enough ("a pinch of this, a dash of that") to make it impossible for you to repeat. The pragmatic rules of language lead to uniqueness and variation, the semantic levels to commonness and repetition.

Uncle Watt had a relatively unique hobby; he collected "fuzzies." He used to take great delight in going to political meetings. Much to the amusement, or distress, of those near him, he would pick out the fuzzies, as he called them, in the politician's speech—the words or sentences which could mean all things to all people. What a heyday he could have with such choice items as the following: "to fight for freedom and light the way to liberty"—"you and I are going to fight for the goodness of our land"—cheers—applause—cheers. Uncle Watt tried to see how many different interpretations were assigned to the same words or sentences. In our terms he was seeing how many different semantic translations were made of a pragmatic statement, or whether it remained at the pragmatic level with the listener simply translating into other pragmatic language to his own liking. Thus, when the politician said, "fight for freedom and light the way to liberty," Uncle Watt would say to one of the listeners, "What's he going to do?" The listener would answer, "He is going to make it possible for a white woman to be free to walk down the street without having to worry about what the black man is going to do, and he's going to make it so that we are free to serve who we want in our stores or our barbershops." To another listener, Uncle Watt would ask, "What's he going to do?" and get the reply, "Why, he's going to make it so that a Negro is free to move out of the black ghetto, free to get an education in a good school, and free to live and work where he chooses." To still another listener Uncle would ask, "What's he going to do?" and the reply would be, "Why, why, why . . . just like he said . . . things are going to be better. Yes siree, things are going to be just fine."

So Uncle Watt had added to his collection of fuzzies a fine new specimen, "to fight for freedom and light the way to liberty," plus the two opposite semantic interpretations, plus one pragmatic retranslation, "things are going to be just fine." "Things" and "we'll see" were two of Uncle Watt's favorite fuzzies. He would say all you need is one fuzzy per sentence and you can get away with saying anything you like. He used to help the kids in their running battle with grownups by instructing them in the artful use of language. He would advise them not to say, "I'll be in at nine," but rather, "I won't be late." Nine is a hard semantic word; "late" is a nice fuzzy which means different things to different people. So while the mother and father argue over whether 9:45 is "late," the heat is off the juvenile language expert. If he had said "nine o'clock" the grownups could then consolidate their forces against the poor culprit.

Logic and Mathematics (Syntactics)

The third level of language important to science is referred to as the syntactic level. Whereas the pragmatic level concerns the rules for attaching words and symbols to personal events (images, thoughts, values), and the semantic level concerns the rules for attaching words or symbols to shareable or tangible objects or events, the syntactic level concerns the rules for combining words or symbols or numbers. This is the language of mathematics and formal logic. At this level of language the symbols need not refer to, stand for, or be tied to anything. They represent nothing but themselves. Thus, a logical statement might read, if A is greater than B, and B is greater than C, then A is greater than C.

Notice that A, B, and C stand for nothing semantic or pragmatic; they are simply marks that differ one from the other, marks which logicians combine and recombine according to certain rules the logicians themselves have invented.

In some cases the syntactic and semantic levels of language can be combined. For example, we might let A, B, and C represent certain concrete objects, such as people, and let "greater than" represent the marks on a tape measure for measuring height. Sure enough, we find that if A, when standing against the tape, covers more marks than B, and B covers more marks than C, then A will cover more marks than C. So this particular syntactical rule holds for at least one part of the semantic world (height).

Notice that we have said that logicians invent these rules, not discover them, and that this particular rule about A, B, and C applies to one part of the semantic world. Look what happens, however, when we let A, B, and C stand for another part of the semantic world, for three football teams: A for the New York Giants, B for the Cleveland Browns, and C for the Green Bay Packers. In this situation we'll let "greater than" stand for winning. Now does it hold that if the Giants beat the Cleveland Browns, and if Cleveland beats Green Bay, of necessity the Giants will beat Green Bay? No, often the top team consistently has trouble with one of the third or fourth level teams in the league. Thus, syntactic rules may or may not hold up when applied to various parts of the semantic world. Some syntactic rules fit certain parts of the semantic universe, but others, such as some mathematical models, have been formulated which probably fit no known aspects of the semantic universe. Mathematical models are man-made inventions, elegant games with complex rules for combining signs and symbols, which may eventually turn out to be useful in describing certain parts of the semantic universe, or which may remain with no known ties to the objective world but stand rather as monuments to man's ability to create and solve complex puzzles with little marks called numbers.

RESEARCHERS USE ALL THREE LANGUAGE LEVELS

All three language levels (pragmatic, semantic, syntactic) are employed in research activity. One way of defining research is in terms of attempts to move from pragmatic and syntactic levels of language usage toward the semantic level.

We can use symbols to describe different types of investigation or research. For example, we can think of any investigation as involving at least two ingredients; in criminal investigations these are an observed crime and a suspect. In scientific investigations the two ingredients are the same: an observed event (O), and a suspect (X) that led to that event. The observed event (O) may be a dead man, and the suspect (X_1) may be his business partner, and tying O and X together will involve a legal investigation. Or the suspect may be cancer (X_2), and so involve a medical investigation. Or the suspect (X_3) may be a new germ, and tying O and X together would involve a bacteriological investigation. But regardless of what type of investigation it is, we deal with observations (O's) and (X's) suspects. And regardless of what type of investigation it is, if you cannot clearly identify your suspects, or if you cannot agree about observations, then attempting to tie suspects to observations becomes not a legitimate semantic game, but a pragmatic game of illusions and suspicions.

LANGUAGE	SUSPECT		OBSERVATION
Syntactic	X	$\longrightarrow$	O
Semantic	Who or what	(led to)	Disappearance of Harvey
Pragmatic	Harvey last seen with?		When?
	Witness 1 says: "A dark man, medium height, glasses."		"Three weeks ago."
	Witness 2 says: "A fair man, tall, thin, no glasses."		"Two or three weeks ago."
	Witness 3 says: "A blond."		"Think I saw him across the street yesterday."

In the above example we are apparently playing our language game on the pragmatic field where we have very little of a shareable nature regarding suspects (X's) or observations (O's). If our aim is to play pragmatics, to daydream, to gossip, to conjecture, to write a novel, then all is well. But if our goal is to bury poor Harvey and hang his murderer, it's nice to have independent agreement as to whom we should bury and whom we should hang. If independent witnesses cannot identify our suspects and share our observations, the ice is too thin for a good game on the semantic pond.

Before proceeding, it is important to introduce some notational aids. If our suspects and observations are tied to objects and events at the semantic level, we can use very solid X's and O's to portray them. If our suspects and observations are tied to images and events at the pragmatic level, we can use different X's and O's. If we are just speaking syntactically and not thinking specifically about what suspects and observations the signs or words relate to, we will use ordinary X's and O's.

We have said that research involves moving from the pragmatic and syntactic levels to the semantic level. Let us start with a syntactic statement: $O_1 + X \rightarrow O_2$, which simply means your first observation will change to a different observation following the introduction of X. Let us imagine the time before penicillin was generally known and before thermometers were well developed. Consider the kind of interplay that might go on between the different language levels. At the pragmatic level, O_1 would stand for fever and X for the new drug, and O_2 for reduction in fever. Our syntactic statement translated into pragmatic terms reads: O_1 (if you have a fever) + X (received new drug) → O_2 (your temperature will fall).

Now to move to the semantic level. Remember that there are no thermometers yet and no standard, well-known procedures for making the new drug, to be eventually known as penicillin. We have a picture, then, of a mad scientist running around pulling at the sleeves of the doctors of the day, saying that he has some green stuff that kills germs and which will likely cure patients suffering from infection and high fevers. We are ready for our first clinical trial.

O_1	X	O_2
Measure fever	New Drug	Measure fever again

Let us look in on the first case. Dr. Pro and Dr. Anti see the patient. Dr. Anti feels the patient's forehead with his hand and says, "Yes, he has a fever." Dr. Pro puts his own forehead against the patient's forehead and says, "Yes, the patient definitely has some fever." Let's assume they do this separately, so we have a certain level of agreement in that both doctors independently put the fever label on the patient. Then they give the patient the new drug, measured out with some measuring method of the day. When they come back the next afternoon, Dr. Anti puts his hand on the patient's head and says, "Maybe yes, maybe no, probably no change in fever." Dr. Pro puts his own forehead next to the patient's forehead and says, "Yes, definitely a drop in fever."

What do we as researchers conclude?

Here we have the classic instance in which independent witnesses cannot agree in their observations. As yet we have no common way of talking about temperature level and changes of temperature. Both O_1 (initial temperature) and O_2 (final temperature) are too pragmatic or too vague.

The first observation, O_1, is not completely vague because both Dr. Pro and Dr. Anti independently conclude that the patient has some fever. However, the terms are still vague enough so that they are unable to agree on the outcome of the test. We need to reduce the vagueness, we need to move closer to the semantic level.

Fortunately, another mad scientist moves into the field. He has invented a glass stick that he says measures fever—all you do is place it under the patient's tongue and see which mark the liquid inside the glass stick rises to. Any physicians with reasonably good eyesight can agree, to within one mark, where the liquid is inside the glass stick. Now we can go back and try to test the new drug again. The glass stick helps us translate pragmatic level language into semantic level language, or at least to move closer to the semantic ideal. With this objectifier (the thermometer) we are now in a position to observe the effects of the new drug on the second patient. Before treatment the physicians agree that there is a fever of 103° or 104°. After treatment they agree that the fever is down to 98° or 99°. Eureka! Success! Next patient—103° or 104° —give drug—wait a bit and take temperature again . . . no appreciable change—wait some more . . . no change . . . no change. What's wrong? Try another three patients—no change.

What can be wrong? The physicians stroke their beards and scratch their heads. One nurse observes that the new drug seems to change color a bit. Aha! Now the question arises as to whether the drug used on the unsuccessful cases (3, 4, and 5) was the same as that used on the successful case (2). Drugs, like milk, change with time. As researchers what do we conclude? It may be that the drug is ineffective and that Case 2 would have improved without any drugs; perhaps we wrongly gave the drug credit. It is also possible, however, that the drug is potentially effective but that it loses its power with time. If this is true, we have a pragmatic or vague drug X in our formula. We cannot agree as yet how to identify the drug. Independent witnesses need some instrument or procedure to help them differentiate, or identify, or label, the drug in its fresh form as opposed to the drug in its deteriorated or sour form. In other words, we need a semantic or shareable way of identifying the drug. So our first mad scientist goes back to the test tubes where he works long and hard at finding measures to detect changes in the drug and ways of keeping it from changing by controlling its temperature and by mixing it with buffers.

Once again he turns to tug at the physician's sleeve. But the physicians are not too enthusiastic. The scientist had gotten their hopes up before only to disappoint them; they don't want anything to do with the drug this time. The heartbroken scientist decides to give up science for technicolor detergents. However, at the next mad scientists' meeting, in a moment of weakness, he tells some of his friends about his new compound and how he made it. They rush home and see if they can make the new compound too. (If they cannot, our friend is probably better off out of the science game.) The other scientists follow the recipe for making the new drug, test it on test-tube bugs, find that it kills the bugs and that it keeps killing them even after several weeks of storage under the conditions prescribed by our friend.

Thus different researchers have been able to repeat his new procedures and all get satisfactory results. These procedures have moved the drug away from the pragmatic and toward the semantic level, just as the development of the thermometer moved the concept of fever away from the pragmatic and closer to the semantic level. These procedures provide us with a shareable way of deciding whether we will label a given drug as drug **X** or not.

Now to test the drug again. Because the physicians in the first hospital are still bitter, the drug is tested in another hospital. What are the results? Case 1—success; Case 2—success; Case 3—success; Case 4—not much effect; Case 5—success; Case 6—success. We now have a semantic level statement, with the vagueness of the key terms ("fever," "drug") reduced to a level that works. That is, the results impress independent experts of the day because the new drug reduces fever faster, or with less effort, than other traditionally used treatments. At the syntactic level we have the statement: $O_1 + X \rightarrow O_2$. The semantic level thus reads: If the temperature is above 99° (measured by a standard thermometer), and if you administer ten units of drug (as manufactured and tested according to Professor Hyde's directions), then very likely the temperature, measured by a standard thermometer will return to normal (98° to 99°) within 24 hours.

You will have noted that there is still some vagueness left; the drug does not work in all cases. Thus, our mad scientist may return to the test tubes to try to reduce the vagueness still further. He may decide that some high temperatures are caused by bugs different from the one his drug kills. He then rewrites the syntactic statement: if $O_1 + O_2 + X \rightarrow O_3$, which, translated to the semantic level, might read: if a patient has a temperature of over 99° (O_1), and if his tissues contain so many type A bugs per c.c. blood (O_2), and if you give compound **X**, very probably his temperature will drop to 98° or 99° (O_3) within 24 hours.

Our scientist succeeds in reducing the vagueness, and research on the problem continues—the challenging and never-ending attempt to describe and predict what happens when you put this semantic bit of nature in contact with that semantic bit—bugs with drugs, radiation with skin, children with T.V., politicians with voters.

Clearly one central concern of science is to increase the degree to which the O's and X's are tied to shareable objects and events, and to decrease the degree to which O's and X's are tied to personal objects and events, and therefore not readily open to test by other researchers.

It is not always immediately apparent whether a given word belongs to the semantic or to the pragmatic level. In fact, probably one of the major weaknesses in the areas of social and medical research is the failure to appreciate this problem deeply, and to be able to make appropriate tests so as to determine the degree of vagueness surrounding the use of the word.

No doubt millions of dollars and thousands of man-hours are spent in research dealing in language at the pragmatic or personal level which is mistakenly used for the semantic or objective language level. Once people become accustomed to using a word, it is extremely difficult for them to appreciate that it may have very little semantic meaning. Because this whole problem of differentiating between pragmatic and semantic levels of language is so critical to research, and because it is so extremely difficult for us to start questioning the semantic base of some of our pet words and phrases, it is useful to have a rough yardstick to help us judge where different words belong on the pragmatic or semantic scales.

LANGUAGE LEVEL YARDSTICK

We have proposed that the researcher attempts to move in the direction of the semantic level of language. This suggests that there are degrees of subjectivity and objectivity, that there are differences in vagueness surrounding the use of different words. Table 1 presents a crude attempt to portray language levels in steps of decreasing vagueness.

Table 1. LANGUAGE LEVEL YARDSTICK

1. Individual Pragmatics X O	Subjective images and sensations for which you have personal labels or names.	I feel gloobic.
2. Group Pragmatics X O	Subjective images and sensations to which you apply labels and words used by others.	"Ghost" "Hell" 50% happy — 20% sad = 30% happy.
3. Potentially Semantic X O	Potentially point-at-able objects and events which require better instrumentation—e.g. better thermometer, microscope, telescope, X-ray, etc.	other side of moon, early signs of cancer, measures of anxiety.
4. Individual Semantics **X O**	Point-at-able objects and events for which you have your own personal names and labels.	a girl is a flurd.

Table 1. LANGUAGE LEVEL YARDSTICK (cont'd)

5. Sub-group Semantics X O	Point-at-able objects and events for which sub-group members use name or label.	a homosexual is a fag; a girl is a bird; a drink is a smash; 25 Ergs.
6. Large Group Semantics X O	Point-at-able objects and events for which various sub-groups use same name or label.	electron pentagon Kremlin 20⁰ Centigrade
7. Sub-group Syntax X O	Rules for combining symbols that do not necessarily refer to any aspect of the semantic world but the rules are familiar only to a specific sub-group.	$(x + y)^2 = x^2 + 2xy + y^2.$
8. Large Group Syntax X O	Rules for combining symbols that do not necessarily refer to any aspect of the semantic world and which are familiar to many sub-groups.	$2 + 4 = 6$

In the case of level 1 and level 2—individual and group pragmatics—we are beyond the reach of research rules. The objects and events involved are personal images and sensations, not available for public inspection. They cannot be pointed at, and so are not appropriate for the research game. They are, however, most appropriate for other levels of discourse, such as talking to oneself or to a loved one, and for use in certain forms of poetic, fraternal, organizational and religious discourse. In the tables, individual and group pragmatics are portrayed as X's and O's to indicate the nonobjective, nonmaterialistic aura surrounding this level of language usage. Language levels 3 to 8 inclusive are appropriate levels of discourse in research. In level 3, the potentially semantic, we work with objects and events that are vague as yet, but offer good promise of being available for public inspection in the foreseeable future as the result of predictable technological development. A classic example, of an object not originally open to public inspection is the other side of the moon. Since the same face of the moon is always orientated toward the earth, no one had ever seen the other side of the moon. However, when technology developed the satellite carrying TV cameras, 'the far side of the moon' changed from level 3 to its position at level 5 or 6, since pictures of it have been viewed by millions. The microscope, telescope, X-ray—all helped to move us from level 3 to levels 4, 5, and 6 with respect to objects and events such as germs, electrons, distant galaxies, and cracks and faults in our own anatomy.

Semmelweis, who predated the discovery of bacteria as a cause of disease, proposed that there was a relationship between death by child-birth fever, and material on the hands of the physicians who had just completed autopsies.[4] In speaking of "childbirth fever," Semmelweis was talking at level 5 (sub–group use of objective language) and at level 3 in talking about "cadaveric" material on the physician's hands. Work by Pasteur in France and by Lister in England proved Semmelweis correct in his hypothesis. However, cadaveric material was translated, with the aid of the microscope, into bacteria, unique objects that could be pointed at and were open to public inspection. Thus we move from a vague level 3, with the aid of individuals like Semmelweis, Pasteur, and Lister, through level 4 and to levels 5 and 6.

Problems

At times it is difficult to determine at which language level an individual or group is operating unless we can carry out certain tests. For example, how can we decide whether an individual is operating at the pragmatic or semantic, or potentially semantic level?

The first condition to be met is that we have a label and a shareable phenomenon. If there is no tangible phenomenon, then we are not yet at the semantic level, as is the case with statements of the type "I feel . . . ," and, "I think . . . ," as well as the more obvious type of statement such as, "I see a ghost." The feelings, thoughts, and ghosts are not available for public viewing, but are the private domain of the individual and, by definition, represent a pragmatic level.

Confusion between the pragmatic and semantic levels very typically arises when you are attempting to label a characteristic of a visible ob-ject—such as bitterness in the cigarette, or thought disorder in a patient. A friend says to you, "I don't see how you can smoke those cigarettes, they taste so bitter." You reply, "I haven't noticed that—are you sure it isn't your imagination?" Thus you are suggesting that the bitterness exists inside his head rather than inside the cigarette. Your friend is not too happy about your suggestion that it is his imagination, and he replies, "Imagination nothing! I would know that cigarette anywhere." You then reply, "Okay, let us do a little test. We'll blindfold you and see if you can pick out my brand of cigarettes from others." You then select ten cigarettes, five of your brand and five of his. You drop them in a hat, shake the hat gently so that the cigarettes are mixed, and have your blindfolded friend reach into the hat and pick out a cigarette, light it, take a puff or two, and tell you what brand it is. You check his guesses against the brand names printed on the cigarette and record the result. At the end of the test you find that out of five cigarettes of your brand

[4] Beveridge, *The Art of Scientific Investigation*, p. 113.

that he tasted, he labeled only two of them correctly. He gets mad at you for making a fool out of him, goes off to tell more people what a bitter brand of cigarettes you smoke, and asks them if they find you difficult to get along with.

What you have done is to test whether your former friend could apply the same label to each of your cigarettes on the basis of taste alone. You know he can apply the label by sight since the label is printed on the cigarette. You demonstrated that he was generally unable to apply the label without having it printed for him on the object. This would be similar to a doctor needing to have the diagnosis painted on the patient's forehead before he could tell the difference between various ailments. This is not as ludicrous as it sounds as we shall see in a moment. In any case, our cigarette example provides an instance of an individual pragmatic use of language (level 1) masquerading as an individual semantic use of language (level 4), and, when put to the test, its appropriate level became apparent.

A similar problem can arise in animal experimentation. A psychologist who is studying the effects of a given drug on activity level in rats calls you in to tell you that the rats injected with the red drug are certainly a lot more active than the rats injected with the blue drug. You look at the rats and cannot see too much in the way of obvious differences. You wonder whether this is the same sort of phenomena as the friend who thought he could tell the difference between cigarettes. You ask the experimenter two questions: (1) Who gave the rats the drugs? and (2) Who measured how active they were? He replies that *he* gave the drugs and that *he* also measured how active the animals were. You say to yourself, "So he knew which animals got the red compound and which got the blue—I wonder how things would work out if he didn't know?" You suggest that you are very interested in the study and you will help him out. However, this time you will give the drugs to the animals without him watching which animal gets which drug; he will then do the test and decide which animals are most active. After he has decided which animals are most active you can tell him how many of them had received the red compound and how many the blue. As it turns out, almost half of the animals that he classed as "most active" had received the blue compound and not the red. This then is the same problem that arose in the example concerning the cigarettes. Without knowing ahead of time which animals received the red drug, the experimenter is no longer able to detect differences in activity level among the rats. Therefore, we conclude either that the drugs do not differ, or that the effects are small and require a more sensitive device for measuring activity level than the naked eye.

Group Pragmatics Are Tricky. We noted previously how

ludicrous it would be if patients had to have their diagnosis painted on their foreheads in order for the doctor to know what the diagnosis was. In many instances, however, this is not so ludicrous as it seems because many observations in medicine and in the social sciences are only at the stage of precision that the observation of fever was at prior to the development of the thermometer; in fact, many medical and social sciences observations are probably even more vague than this.

Let us start with a very simple proposition at the pragmatic level. If someone has a weak ego (O_1), and if they are stressed (X), then they will develop neurotic behavior (O_2). Let us first attempt to decide whether the term "weak ego" can be applied consistently by independent witnesses or observers. We place different judges, or different doctors, in different rooms. Then we take our first patient, Oscar, to each judge independently. The judge has the job of attaching one of two labels to Oscar—the "weak ego" label or the "not weak ego" label. No judge is allowed to know what the other judges have decided prior to making his own decision.

If the independent judges always agree which label goes with which patient, then we have met the minimum rules for playing at the semantic level. If the judges do not agree, then we are at the pragmatic level of language in which each judge is playing by his own personal rules of assigning labels to objects. We can have still another case in which not all the judges agree; but if even some agree consistently, then we have some judges playing by semantic rules (level 5) and others by pragmatic rules (level 2).

You will notice we stressed that each judge was to make his decision alone, away from the other judges. People can agree with each other for various reasons. One is that they are independently using a common yardstick. If, however, they have seen the patient together, another reason for their agreement might be that they are playing follow the leader. It is critically important to be clear about the difference between these two reasons for agreement. For it is this difference that separates research, that plays by semantic rules, from pseudo-research, that pretends to play by semantic rules. When your smoking friend must peek at the label on the cigarette, or when the doctor must read another doctor's notes to see his diagnosis, then you have an area of vagueness surrounding the use of that particular label. The follow the leader procedure is fine as a training technique to move from level 4 (individual semantics) to level 5 (sub-group semantics). But the test of when level 5 has been reached must be agreement between independent observers in assigning a label. Otherwise you have no way of differentiating level 5 from level 2 (group pragmatics) or the mob effect. There is probably no distinction that is more important than this in differentiating science from nonscience. And yet, because making such distinctions is usually laborious and frequently threatening, it remains too often an ideal, rather than an inviolate practice.

SUMMARY

We have discovered that you speak at least three languages—pragmatics, the language of love, religion and poetry; semantics, the nuts-and-bolts language of shareable, point-at-able things; and syntactics, the pure language of symbol games like mathematics. There is no supreme court ruling on language usage; we rely on all three languages, now emphasizing one, and now another, but usually mixing pragmatics and semantics into a general–purpose language stew. The first rules of the research game are to know how to tell one language from another, and to attempt to make your major plays in the semantic field where independent observers can share your key observations, can identify your prime suspects. If your observations are vague or private and your suspects melt away under the scrutiny of others, you play the pragmatic game, or one that wanders back and forth between the semantic and pragmatic fields. Here the naive spectator becomes confused or hoodwinked. Furthermore, if either your key observations or suspects are vague, it is usually difficult to play the science game in a first-division pro league for any length of time. But you can still play in exhibition games or in the minor leagues. And you can still play the scientist on TV interview shows or with people who don't know the rules, or who haven't seen you play.

Thus, the science game can be distinguished from nonscience games in terms of the emphasis placed on certain language rules. Another distinction commonly made between science and other activities concerns what constitutes acceptable evidence. We will examine this topic in the next chapter.

Rules
of Evidence,
or Who's Right

Regardless of which language we use—the language of nuts and bolts, of mathematics, of religion—we are almost continuously wrestling or playing with the following questions:

What do you know?
What do you know for sure?
Who's right?
How do you know?

TWO APPROACHES TO KNOWLEDGE

These are questions we will labor over and fight about with at least the same commitment as we would over a mate. These are questions of knowledge, decision, belief. How shall we approach them? We can use the T-F approach, or the D-M approach, or perhaps a combination.

True-False Approach

The T-F, or true-false, approach is built on the idea that there are packages of true knowledge and packages of false knowledge, and that man's job is to be able to tell the difference between them. Sometimes this is an easy task:

> I smell skunk!
> We smell skunk!
> We run!

As long as the data immediately available is clear, loud, hard, or strong —like elephants, bombs, rocks, and skunks—most people don't seem to need any fancy theories about "how we know" or "who is right." Using the notational aids developed in the last chapter, we could portray this kind of data as **X**. In the case of the skunk, the rare person with an unusually bad cold, his back turned, and no friends within shouting distance may get caught with a spray. But surely we don't require an elaborate theory of knowledge to handle these extraordinary instances. Rarely do we get into arguments about the current presence or absence of skunks, freight trains, or explosions. With these well-organized data packages, the true-false approach seems to work well enough. People who don't agree with the majority on these kinds of questions we send to a psychiatrist, or we cut them off the bottle, or we take away their grass.

The T-F approach serves us well in dealing with 'loud and clears' under our noses, but even though we use it, it serves us less well when we attempt to package the past, the future, or even the present when it involves what's happening around the corner, in the mind, or when the signals are weak, or many, or changing fast. How can we decide what is true and what is false when most of the data is beyond our reach? Perhaps we need an alternative approach.

Decision-Making Approach

The decision-making approach to the question "What do you know?" goes something like this: with the information you have immediately available, with the figuring skills and habits you possess, and with the time and energy you can muster, you choose this package of knowledge over that one—at least for the time being. So, according to this view, what we know at any given time is a blend of what we've just noticed and what we've remembered, combined according to a recipe we have the skill, the time, and the energy to use.

If we would know more than skunks, if we would draw conclusions and say wise things about data bits scattered widely through time and space, then concepts like 'true' and 'false' are best replaced by concepts like 'acceptable' and 'unacceptable.' And it appears that what we most want to know concerns just such data bits scattered widely through time and space—we want to know about Republicans and Democrats, cancers and cures, teachers, teenagers, pollution, black men, T.V. stars. We want to know about distant things like Paris, the moon, and Judy's university. We want to know about what's going on in hidden places like bedrooms and back seats, what's in the coach's mind, in the girl's heart, in the boss's private file. We want to know about 20 years from now and about 20 million years ago. And so as individuals, as groups, as institutions, we develop data-sampling and data-chunking rituals to make up our knowledge packages because we have no other choice. Like you, like the press, like the courts, science represents a pool of data-sampling and data-chunking rituals. But, like your conclusions, the conclusions of science are manufactured with the bits of data and the data-mixing rituals current talent and time permit.

Decision Aids. Man, the believer, the decision maker, continuously must make choices, often on the basis of little or no current data. He uses, out of desperation, or habit, or boredom, or exhaustion, whatever decision aids he can—anything that prepackages information and presents it in a form that can be shaped with the talent, time, and energy at our disposal to fit some corner of our mind. Our decision aids include a wondrous assortment: whispered washroom rumors, a jury's verdict, the *Reader's Digest*, the committee vote, a chairman's ruling, a confidence overheard at the pub, the radio report, the newspaper editorial, the football hero's prescription for breakfast food, the president's assurance, a glimpse through a keyhole, a half day at the library, a panel discussion, the flip of a coin, the consultant's recommendation, the man-on-the-street survey, a national poll, a pulpit pronouncement, a race tout's tip, a stockbroker's recommendation, the old lawyer's advice, the young doctor's opinion, a professor's view, a friend's assurance, the salesman's figures, the mechanic's decree, the party's policy, the mob's choice, the customer's complaint, the witness's report, the student's protest, and the scientific study. Surprising, isn't it, how rarely we have access to raw data? How frequently we rely on prepackaged data without being aware of how it was collected or shaped, but ready to call it true or false.

In the next section we will examine some of the data-collecting and data-sorting methods and decision aids currently employed in science. But first let's look at some of our decision aids, at our preferred rituals for collecting and chunking data—rituals and aids we use regularly in helping us decide what to accept or to reject, what to label true or false.

HOW DO YOU KNOW?

Consider the following examples and try to decide how you would judge for yourself the acceptability of these propositions. Dad says, "Never trust a college professor—they're all a bunch of radicals or softheads." The television commercial says that Cheer washes are whiter and brighter. A newspaper reports that U.S. losses in Vietnam are light. The minister preaches in his sermon that we are all God's children. An Olympic runner announces, "Smoking smothers speed so I don't smoke." A journal contains an experiment, the findings of which indicate that a well-known toothpaste reduces cavities. An anthropological report suggests that the Piltdown man is the missing link.

An examination of these examples will indicate that it is not possible for you to assess the truth or falsehood of a general claim on the basis of your personal experience or observation; even if it was, since your time and energy are limited you want to apply your limited knowledge-seeking resources to areas that are of cardinal interest to you. Nevertheless, there are many important situations when a decision is required and we are unable to make personal observations relevant to the issue in question. Or perhaps the matter is extremely complex and, although we have had some personal experience with the issue, we know of large gaps in our data. What do people typically do in these sorts of situations? How do they decide what to accept or what to reject, what to call true or false?

Bias Boots

Often bias (prejudiced consideration of an issue) is used as a kind of evidence in the assessment of truth or falsehood. And there are many varieties of bias and many ways in which biases may be expressed. We can think of bias as a pair of boots—different biases are represented as different styles and colors of boots.

Cultural Biases. We are probably most familiar with thinking of biases as the property of individuals; however, it should be noted that there are cultural biases as well as individual biases. Cultural biases represent instances of large numbers of people wearing the same color and style of bias boots. Such group biases are called "folk truths" by the sociologists, and "conventional wisdom," a term coined by economist John Kenneth Galbraith, is also appropriate.[1] As an example of folk truths, consider some of the old saws, axioms, or sayings of our culture:

[1] John Kenneth Galbraith, *The Affluent Society* (Toronto: The New American Library of Canada Limited, 1958), Chap. 11.

"Look before you leap" or "He who hesitates is lost." Most of us at one time or another have found ourselves spouting one of these trite adages when we give advice or when we're frantically searching for something to support our arguments. Such sayings are usually harmless, but are also impossible to prove or disprove since the conditions under which the statements should apply are never specified. Also, they are always right and always wrong since, as in the above example, they are usually contradictory. Even though they grossly oversimplify the data, they do serve as decision aids or decision justifications.

Common Sense. We also have biases or points of view expressed in terms of good old common sense. Some appeals to common sense are based on acceptable technical knowledge, but the individual is unable to produce the technical details and therefore justifies his argument by stating that "It's just common sense." Such is the case of the woman who pulls out the plug of her television set during an electrical storm because it's just common sense, even though she is not even remotely aware of what specific disaster she is avoiding.

Other appeals to common sense are merely personal opinions parading behind the common sense front. For example, some people argue that it's just common sense that capital punishment be the penalty for murder. In many cases they have little durable evidence to substantiate their contention; but they often neglect to present it as a hypothesis or opinion. They make the statement as if it were well established and label it as common sense to increase its acceptability—and if you don't concur with the statement, there's a not-so-subtle implication that you're stupid.

Self-evident Truths. There are other labels like common sense that are used as data props or fronts. Consider a famous portion of the American Constitution. Thomas Jefferson and his cosigners stated in 1776, "We hold these truths to be self-evident, that all men are created equal, that they are endowed by their creator with certain unalienable Rights, that among these are Life, Liberty and the pursuit of Happiness." The term important for our purposes here is "self-evident," which means that what is described is evident in itself without proof; it is axiomatic. This approach is akin to what some philosophers and laymen have called a priori knowledge, information which is assumed to be true without experience. In other words, we know it is right because our intuition tells us so. In the section from the American Constitution there is no attempt to offer any evidence from experience or observation that all men are created equal (indeed, a lot of data would seem to indicate just the opposite), but the statement is formulated in such a way as to discourage questioning.

Bias—Pros and Cons. Bias, although serving a vital function in protecting us from decision overload, is often very resistant to change

—that is both its strength and its weakness. However, most biases do fluctuate in strength, suggesting that under some conditions at least, they are amenable to modification. Most of us are aware of some of our biases even though such awareness produces little obvious change in our behavior. The well-educated Arab, in his calmer moments, may be fully aware of the futility of attempting to regain all of Israel. Nevertheless, this awareness does not counteract his anger at what he sees as injustice and unlawfulness in the handling of the Israeli situation by the British and the United Nations.

Since bias often has a strong emotional component, this acts to make the bias boots durable and thick-soled, protecting the wearer from painful experiences as he marches through life. If believing something to be true will help us in some way, we, understandably, often accept it as true without careful examination of contrary evidence. Whether it is a bitter racist, or a petty cheater on income tax, or a husband shopping for dessert, bias steers minds into a decision and then helps justify it after the event.

While biases simplify our decision making, at the same time they reduce our learning capacity. Because biases are tightly organized, well-wrapped chunks of data, they are difficult to penetrate. Information which is not in accord with our biases is usually ignored, or perceived in a distorted way. The arguments of Semmelweis and Harvey were not sharp enough to pierce and deflate the biases of their colleagues. If new information is congruent with our old beliefs or biases, we welcome it as an old friend. But if the new information cannot be readily absorbed into a tightly organized existing belief matrix, it is rejected out of hand, or shaped to fit, or nailed on the rear as the rare exception that proves the general rule. Thus biases help us hear what we want to hear and see what we want to see.

Discussions in the cocktail lounge, over the bridge table, at the fights, in the political arena—generally the idle chatter in which we engage—is peppered with bias. Sometimes we are aware that one of our remarks may appear to lack complete objective validation, and we graciously preface our pronouncements with "I think" or "In my opinion." At other times there is a delirious absence of any awareness that we are wearing our heavy bias boots. Often in such discussions the loud and aggressive speaker is able to erode away, or at least cork, contrary views.

Origins. A word about the origin of biases—we learn our biases in the same manner and from the same sources that we learn reading, writing, spelling, football, fear of the dark, and table manners. Sometimes we learn them so well they seem inherent—as normal as five fingers and two eyes. We have no difficulty identifying the biases of others, but often are blind to our own. Without biases, however, we would be swamped by a myriad of alternative views, so maybe we need contests to see who holds the most beautiful and least damaging biases.

Appeal to Authority

Still another type of evidence used over the dinner table, in the pub, and in committee meetings, to settle the issue of what to accept and what to reject is an appeal to authority. If an authority supports the proposition, then this is often regarded as sufficient evidence.

In early childhood parents are our greatest authorities, but as we get bigger, they get smaller, and we rely on prestigious institutions like the Church, the government, science, and students a year older than we are to prepackage for us data ranging from politics to contraceptives.

Can you think of other institutions that we allow to collect and chunk important data for us? In addition to television and the press, an obvious example is our entire legal machinery—policemen, judges, prosecutors, juries, legislators. The state, in the form of the prosecutor, says the suspect is guilty of murder while the suspect, in the form of the defense attorney, says, "I'm innocent." Who is right? Twelve reasonable men and women consider the evidence presented in court. The prosecution collects, cuts, and shapes data to fit the guilty bias; the defence collects and shapes the data to fit the innocent bias; and the jury decides which did the most convincing job. We, the public, delegate to the courts this task of data collection and analysis, and, furthermore, we usually regard the outcome as being an acceptable one.

The French courts decided that Alfred Dreyfus was guilty of treason and convicted him in 1894. But Emile Zola wrote an open letter about the case and eventually, after a second trial in 1906, Dreyfus was proved innocent.[2] He was *proved* guilty in 1894, but *proved* innocent in 1906. Isn't it amazing what time can do to the truth? It's not so amazing, however, what time can do to the data collecting, cutting and fitting rituals by which truth is manufactured.

The point is that we must delegate much of the task of data collecting and chunking to others whether we like it or not. To an astonishing degree we must rely on the observations and data analysis rituals of individuals, living or dead, and of institutions, near and far. If we have to rely on authorities, how do we choose a good one? Why, pick an expert, of course!

Appeal to the Expert

An expert is a high-priced authority, an authority with credentials. When we are not sure whether some particular point of view is

[2] Emile Zola, *J'accuse La Vérité en Marche* (The Netherlands: éditions Fasquelle, 1901).

acceptable, a seemingly rational approach is to look to the expert. But one of the big problems is choosing your expert.

In order to be considered an expert it seems that the basic qualification is notoriety. For example, Bertrand Russell's stating that banning both underground and atmospheric tests of nuclear weapons will promote the cause of peace is much more impressive than if Gracie Hoe made a similar pronouncement. So, in order for our appeal to the expert to have a chance, we must choose as our expert an individual who is reasonably well-known.

Having found a relatively famous person, will we then accept him as an expert? The chances of our acceptance are increased if the expert's point of view on the issue in question happens to agree with our own. If you are neutral, the expert has a good chance of convincing you; and he still has a fighting chance if you are mildly negative. If you are strongly negative, it's probably a waste of time for others to appeal to experts in their efforts to convince you.

Suppose we are lucky enough to find a famous person whose views on a particular issue agree with our own. Does this automatically make him an expert? Some politicians have advocated the use of nuclear weapons in Vietnam and this may seem like a sensible course of action. Are these people experts? "Of course not!" you may reply, "What do they know about *military* problems in Vietnam?" Obviously, before we regard someone as an expert, it's nice if he has had training or broad experience with the data in question—that sounds reasonable. Then why are some columnists experts on all of life's problems? Why are certain newscasters regarded with reverence and awe? Why are some religious leaders experts on American foreign policy? Why are the military leaders experts on American domestic problems? Maybe our selection of an expert is not based on how much he knows, but whether he can help us make a decision—good, bad, or indifferent.

Experts Reconsidered. The major difficulty with expert evidence is that it is indirect. In every instance we are relying on data collected and chunked by others. We haven't investigated the issues personally, but we assume that those we learn our biases from, our authorities or experts, have made adequate observations and done competent analysis themselves. We are, in essence, relying on hearsay evidence, and we don't know how our experts arrived at their conclusions —in many cases we can assume that they got them from someone else. Well, anything's better than not having an opinion!

As noted previously, one of the challenges in depending on experts or authorities to determine what is acceptable is that experts often disagree among themselves. Light consists of waves—no, it consists of particles. Nicotinic acid can cure schizophrenia—no it can't, but psychotherapy can.

One garage mechanic says that Chrysler products are the best, another says that GM cars are far superior. One economist reports that recessions can be avoided by altering banking policy, while another states emphatically that recessions are inevitable in a competitive free enterprise system. God exists—God is dead. One scientist writes that birth control solves the problem of the population explosion outstripping the food supply, while others stress improved technology in the production of food as the solution to the starvation problem. One group of doctors advises a fat free diet, another group says it is nonsense. One critic pans a Broadway play as the worst ever written, while another reports that a better play has yet to be performed.

Even when experts do agree, you can't be sure they won't change their minds. Some years ago mentally ill people were regarded as being possessed by devils. The cure was to beat them and, if this didn't work, to chain them up. Most experts, with the notable exception of the French physician Pinel, agreed that this procedure was the most effective.

So experts, like biases, present us with a dilemma: it's hard to live with them, but without them each of us would face a horrendous problem of decision overload. So we trust those experts who don't offend any of our deeply ingrained biases.

Comfort in Numbers

Biases and experts are both useful and necessary decision aids, but nothing is more exhilarating and reassuring than group agreement. But if the group goes against you, is there anything more depressing and deflating, or, at times, more frightening? Whether we like it or not, most of us accept majority expression or majority vote regardless of whether it is a majority of one or thousands. Majority vote is potent; unanimous vote is lethal—it indicates fatigue, fear, indifference, or the presence of a skunk.

Large group agreement has supported such data packages as: "There's a skunk around here." "The world is flat." "We're going to lynch Horace." Once again we see the T-F approach has obvious limitations. Large group agreement, however, usually assures good durability for the data package. The-world-is-flat package was around for a long time and, for most of us, is still personally valid. The data samples we get, except for the odd shot from space or the book store globes, continuously support the flat package. So, with the data readily accessible, the flat view makes sense, and contrary data doesn't fit in with a whole host of biases.

Large group agreement does appear to be an effective method of assuring the durability of a data package (unless the large group is a mob). When large group agreement represents agreement among widely scattered people it can mean that (1) we are dealing with a clear-cut data sample like skunks and flats, or (2) we are dealing with a vague data sample, but a strong cultural belief or bias (religion, politics) which is deeply ingrained. In either case, changes in the data package will usually be slow.

When group agreement represents agreement among members of a mob, we are dealing with vague data samples and very quick and dirty data sorting, weighing, and combining procedures. Such a hastily concocted package has all the immutability and power of a jar of nitro.

Again the dilemma—group agreement is a powerful and common decision aid with great potential for order and for disorder, for stability, but also for rigidity.

One in Infinity

From birth to death, people are great snoops and Missourians all. With the talents and energy at our disposal we are supremely nosy —we look, we glance, we stare—we listen, we eavesdrop—we read and dig and touch and smell and argue endlessly—some in public debate, some of us in small back rooms, and some in the quiet or frenzy of our own minds.

In this age of experts and polls it is perhaps wise to recognize that the individual remains unique in spite of our attempts to prepackage everything from pablum to point of view. Each of us is guaranteed our individuality—for good or ill. And so a final decision aid is your unique point of view, your unique data samples that color your decisions. Among the blur of humanity your uniqueness may not be apparent, but each of us arrives at our decisions by a somewhat different route. Each of us questions and shapes—each of us is a creative artist or a kook, depending on who does the labeling.

WHY USE THESE DECISION AIDS

All these decision aids (biases, experts, authorities, group agreement, unique personal viewpoint) are familiar. None can guarantee packages of data that will remain immune to the erosion of time, or that will prove accommodating to additional data, or that will be impervious to the cutting tools of other analysts. This being the case, why do we use these common types of evidence so frequently?

Who Cares?

One reason for relying on such evidence is that we often have little or nothing to lose. In some cases, whether or not we make one choice or another is unimportant because the risk associated with making an error is small. For example, many of us don't care one way or the other whether Cheer gives a brighter wash. It really doesn't matter too much what soap you use—the difference in the cleanliness of the clothes after washing with different detergents is likely nonexistent or very small. Not only is the risk of error small, but many of us are also just not interested in the issue and are therefore unwilling to devote the time and energy necessary in trying to establish which, if any, soap gives the whiter data. And so we act on habit, or we buy the cheapest product available, or the one we remember, or recognize.

Low Computing Costs

Another reason for using these common decision aids is that the computing costs for this type of evidence are low. (By computing cost we mean the time and effort required to collect, sort, and weigh information.) We sometimes make important decisions with such aids because the computing costs to obtain other kinds of evidence or additional data are too high. Perhaps we haven't the time or energy—we have to make a decision tomorrow; perhaps we haven't the money—in order to get the data we'd have to hire a staff of fifty for two years to gather it and rent a computer for a month to analyse it; perhaps we lack the necessary knowledge or experience to investigate the issue. In any case, there are many instances when vital decisions must be made on the basis of prepackaged evidence because of prohibitive computing costs.

No Choice

A third reason, a corollary of low computing costs, for using such evidence, is that we have no choice. It is physically impossible to verify all the information we deal with in an hour, let alone a day. The computing costs in terms of time, money, manpower, equipment, and facilities would be enormous, to say nothing of the fact that our eternally questioning attitude would drive us, as well as our friends, crazy. And so we are forced to accept information, or reject it, at face value. If it fits in with already established biases and data samples, we accept it. In terms of our day-to-day living, there is no more frequently used method for assessing truth than comparison with our biases.

DIFFERENCES AMONG DECISION AIDS

We find some types of decision aids more acceptable than others. Surely the jury's verdict warrants more confidence than the toss of a coin. Why? Is it because the former is more likely to be *true?* Or is it that one is less open to gross attack? In the case of the jury, the data and the data chunks have been arranged and rearranged until twelve different minds can all agree. So it may be that in many areas of discourse we will shift away from concepts of true and false and shift rather towards those data collecting and processing procedures which produce relatively durable packages of information at a cost we are prepared to pay.

Increasing the Durability of Packages

We can increase the durability of a package of information by increasing our data collection and processing costs. We can consult not one, but several, experts; however, this takes more time and money. We can read more about the topic, but this takes time—and it's difficult to understand some of what we read. We can take the car to several garages, but then which one do we believe? We can take a course on automobile repairs so we'll be in a better position to know if the mechanic is cheating us, but think of the time and the effort. We can ask the expert, or authority, or friend to explain in detail his data, his data-sorting, weighing and combining procedures. This takes time and often we can't understand unless we duplicate his training—unless we learn the language and biases of the expert. We could demand that no man be declared guilty until twenty-four, instead of twelve, reasonable men agree on the verdict. Or we could stipulate that before being declared guilty the accused must be convicted in two separate trials with different judges, juries, prosecutors, and defence lawyers. But all these procedures, while perhaps increasing the durability of the decision or data package, would increase our decision costs. What we are doing is providing increased opportunity for additional data to be added, allowing additional opportunity for alternative ways to sort, weigh, and combine the data, and demanding that it be acceptable to a larger number of our representatives. We are increasing the durability of the decision, but at terrific cost, and always knowing that not all relevant data has been gathered, always knowing that alternative decisions remain.

Going All Out

In the main, we seem to protect our limited data-gathering and analysing resources for a few areas that have high personal priority —baseball, dancing, cooking, war. In these few high priority areas we like to gather some of our own data—often very carefully and sometimes in large amounts. We take great pains sorting it, weighing, arranging it in different ways until we find a way that satisfies us, and perhaps satisfies a few others. The data we collect and arrange and rearrange may be musical notes, flowers, dance steps, words, football statistics, germs, or electrons. Some of these activities we call science, some we call art, some we call hobbies—but all of them involve careful data collecting and arranging and rearranging until we find an arrangement that we can accept—for the time being.

In the next section we will examine some of the rituals for gathering and sorting data that are currently preferred in the science game.

two

Sieves
of Science

FEW of man's tools can surpass in importance those used to sift fact from fancy. We will now consider the most powerful techniques evolved to date in our attempts to unmask fact and fancy, those two delightful knaves who love to masquerade in each other's costumes. These two masters of disguise are embraced with vigor by scientist and nonscientist alike. At unmasking time, scientists are often caught, much to their embarrassment, clutching an armful of charming fancy instead of beautiful facts.

In the preceding chapter we discussed some of the rituals and decision aids which all of us use in deciding at any given time which packages of data to accept and which to reject. In this section, we will concentrate on the techniques evolved by scientists for this purpose. All of these techniques are extensions of some of the forms of evidence-finding procedures discussed previously, but frequently these extensions involve more elaborate data collection and analysis procedures.

It should be emphasized that the methods which men of science have developed to sift fact from fancy are not perfect. They are not expected to eliminate error. Rather, they are designed to increase the durability of the data packages produced by science. Some of the methods produce more durable packages than others. The different methods scientists use may be thought of as different sieves; some of them have been designed to collect large chunks of data, others to collect small chunks. Some sieves are designed to collect relatively durable chunks of data, large or small, and others collect an indistinguishable mixture of perishable and durable data. Usually, the larger the chunk, and the more durable you want the chunk to be, the greater the cost involved in designing and using the appropriate sieve or sieves.

chapter **4**

Naturalistic Observation

Watching the baboons at a zoo briefly, it might seem, to judge from the number of fights occurring, that these animals are extremely ferocious and savage. Is it valid to draw this conclusion? Are we justified in saying that baboons are aggressive? Suppose we spent only a few minutes watching them in their rather cramped quarters while, at the same time, we were wiping ice cream off Johnny's face and calling to Jane to stay away from the fence. Our observations of the baboons were casual, to say the least. In 1960, when baboons were studied in their natural habitat, where they could roam freely, the researcher reported little violent behavior. The method of science used in this instance was naturalistic observation: the essence of the method is the observation and description of events as they occur in nature or naturally. There is no manipulation of events, no controlled experimentation, but merely the observation of events as they occur in their usual surroundings. In some scientific circles, the technique of crude observation of events (without highly technical apparatus or instrumentation) is known as the "eyeball technique."

Imagine, for a moment, a world where everything is totally strange or unfamiliar. In such a world, each event is a unique happening and, to the naive observer, is unconnected or unrelated to anything that has gone before or will come after. There are no general laws, no underlying principles, no rhyme or reason to things as far as our naive observer is concerned. This may have been the position of the early man-apes, *Australopithecus*, at the dawn of their awareness. Initially there must have been surprise, fear, and consternation when the colors of leaves changed in the fall and the temperature dropped in the northern climates. Gradually, however, this change from summer to winter came to be expected and one winter was seen to be similar to other winters in certain respects. And so early man perceived pattern, regularity, order, and this discovery or invention of pattern is the basis of science.

Picture the shipwrecked astronaut many years hence as he lands not on Mars, about which we already have much information, but on some totally unknown planet. Will he begin by experimentally manipulating events? Will he set up a tub of water and immerse objects in it to test the operation of Archimedes' principle on this far-flung planet? Not likely! In all probability, he will try to find out where he is and to read what information he has available on this planet. Suppose he has none or can't determine where in the universe he is. He will explore and reconnoiter his surroundings—he will observe the events on this planet as they occur naturally. He will look for similarities and differences. He will classify such events. He is, in essence, looking for general principles or laws which will enable him to make some predictions about what events can be dangerous for him.

The same general principle is applied in some of our oldest maxims— "You must walk before you can run," or "Look before you leap." The implication is that the fundamentals must be grasped before sophisticated techniques will work successfully. The basketball coach does not begin by teaching complicated plays. First he trains the team on the fundamentals—dribbling, set shots, and lay-ups. Once these have been grasped, they are then combined in complex ways to produce highly intricate plays. So it was with the history of science: the primitive method of observation and classification was used before the complicated experimental methods were employed with the aid, in many cases, of elaborate instrumentation or apparatus. And even with these advanced methods, skills in techniques of observation are essential.

There are perhaps four aspects to the method of naturalistic observation as now used: (1) detailed observation, (2) accurate recording, (3) non-participant observing, and (4) grouping or classification. Darwin's exposition on evolution provides an excellent example of all four elements at work. Darwin's description of his own activities will serve to portray in part the operation of this method.

When on board H.M.S. *Beagle* as naturalist, I was much struck with certain facts in the distribution of organic beings inhabiting South America, and in the geological relations of the present to the past inhabitants of that continent. . . . On my return home, it occurred to me, in 1837 that something might be made out on this question [of the origin of species] by patiently accumulating and reflecting on all sorts of facts which could possibly have any bearing on it. After five years' work I allowed myself to speculate on the subject and drew up some short notes; these I enlarged in 1844 into a sketch of the conclusions which then seemed to me probable: from that period to the present day I have steadily pursued the same object.[1]

Thus Darwin, by his own admission, devoted five years to careful, detailed observation of plants and animals, both domestic and wild. Ruminating on his observations, he grouped events, i.e., he classified or ordered what he had observed into categories and, as a result, developed his theory of evolution. If we were to present this schematically we would start with a series of O's (Observations) and then impose some ordering scheme on our O's so that they are combined according to certain rules.

$$O_1, O_2, O_3, O_4, O_5, O_6, O_7, O_8, O_9, \ldots, O_n{}^2$$

then

(1) $O_1 = O_2$ in some respects
(2) $O_1 \neq O_2$ in some respects
(3) $O_3 < O_4$ in some respects
(4) $O_3 > O_4$ in some respects

We could translate this grouping or classificatory scheme as follows:

(1) The kittiwake gull is like the herring gull in that both employ a choking posture which advertises the nest site.[3]
(2) The kittiwake gull is unlike the herring gull in that the herring gull uses

[1] Charles Darwin, *The Origin of Species* (London: J. M. Dent & Sons Ltd., 1928) p. 17.
[2] Each O stands for a different observation. If there were 19 observations then n would be 19 and O_n would represent the 19th observation.
[3] Konrad A. Lorenz, "The Evolution of Behavior," *Scientific American*, CIC, No. 6 (December, 1958), 66–78.

the oblique and long call postures for defense, while the kittiwake uses the choking posture.

(3) The chimpanzee is superior to man in the respect that he can run faster than man.[4]

(4) Man is superior to the chimpanzee in terms of the distance man can cover. He can walk for many miles.

In the discussion to follow, we will examine each aspect of the naturalistic observation method in turn. In actual practice, however, these facets of the method do not necessarily operate sequentially or independently.

DETAILED OBSERVATION

Distortion

Personal experience or observation of an event is not a guarantee of truth. Since human senses are fallible, what we think we see isn't always what has occurred. Our observations are not pure. That is, we do not perceive only forms, contours, certain wave lengths of light or sound, but we also impose an organization on them. One might say with some justification that we *do not* in fact see light waves of 75 microns, we see red. Similarly, we do not hear a sound of 80 decibels, rather we hear a pneumatic drill. Thus any event that is observed is not experienced in the raw, so to speak, but is altered by our past learning. Sometimes our interpretations (organization) of what we see can be quite misleading. For example, psychologists have constructed rooms built on a slant in which an individual standing at one side of the room looks like a giant while the same individual on the other side of the room looks like a midget. Instead of perceiving the room as distorted, we distort the size of the individual in the room.[5] If we press this argument further, it will be seen that our opinions, beliefs, and attitudes can also alter our observations, since they form part of our past learning. Now, when we are observing the movement of planets, the pelvic bone structure of the apes, the strength of a magnetic field, or the electrical conducting properties of copper, you might argue that political affiliations, religious denomination, skin color, or nationality would not affect our observations. Is this, in fact, the case? Religious beliefs supporting the theory that the earth was flat prevented many men from making the simple observation that the mast of a ship

[4] Sherwood L. Washburn, "Tools and Human Evolution," *Scientific American,* CCIII, No. 3 (September, 1960), 62–75.

[5] M. Weiner, "Perceptual Development in a Distorted Room: A Phenomenological Study," *Psychological Monographs,* LXX, No. 423 (1956).

appeared on the horizon before the rest of the ship. And, if the area under study involves human behavior, how much greater will be the distortion of observation!

The scientist using the method of naturalistic observation strives to be as objective as possible; he consciously tries to observe without judging. He tries not to make value judgments like good or bad, wrong or right, beautiful or ugly. In other words, he attempts to prevent his own biases, opinions, values, and beliefs from coloring his observations. He tries to keep observations as pure as possible. This is, in fact, an impossible task that the scientist has set for himself, but at least he is aware of the problem and can caution himself against it. He has, in many cases, also received some training along these lines.

Selection

We pointed out earlier that biases play a role in the selection of information. In other words, we see or notice those things we want to see, and screen out information that doesn't fit in with our particular point of view. For example, the biased observer who is convinced that civilization corrupts man, and who believes that men living in primitive societies are happier, may fail to notice all the negative aspects of primitive life. The poverty, the suffering that is due to lack of medical attention, the gruelling hard work with improvised tools—these things escape his attention. The scientist is aware of this pitfall and so makes *detailed* observations.

Reducing Error

Making several observations and having several observers make repeated observations of the same event increases the probability of producing durable packages of information. This principle is utilized by those practicing the method of naturalistic observaton. In an attempt to partially overcome or reduce observer bias, scientific investigations often make use of two or more independent observers who later come together and discard any observations on which they do not agree. By the same token, the same observer may try to observe the same event many times in order to ensure that he has noticed all the relevant details and to rule out the possibility that his initial observation was a once-in-a-lifetime event. Thus, anthropologists may visit the same primitive tribe many times, and observe their activities over long periods. For the same reason, an experiment may be repeated by another investigator and, to the extent their findings agree, their observations gain in durability.

Another increasingly popular method of obtaining accurate observations involves the use of recording equipment, such as video tape recorders or movie cameras. There is still the danger that valuable data may be lost, but the observations obtained are not so subject to distortion, decay, or growth as are the ones we deposit in memory. A further advantage is that the data may be perused in all its rich detail at some later date and may be scanned again and again, by you, or by others.

It should not be assumed that the problem of accurate observation is a problem only with the method of naturalistic observation. It is a problem with any of the methods of science in which, in the last analysis, the measuring instrument is the human organism. The scientist, however, because of his training and use of the safeguards described above, may be in a better position to collect and store his data than the non-scientist.

ACCURATE RECORDING

Having made what we regard to be careful and detailed observations, we are now faced with the problem of recording these observations. For a scientist, it is not enough that he satisfies himself about the adequacy of his observations; he wants to share his findings with his colleagues and convince them. To do this, he must convincingly record what he has discovered. Scores of books are available on the mechanics of good report writing, and it is not our intent to advise the reader on such matters except in a very general sense.[6] Our concern is more with what is reported than how it is reported.

Selection Through Erasure

Just as selection operates on our observations, so does it operate in our report writing. We don't see everything that happens, nor do we record everything we see. We can't record all our observations (even if we wanted to) because we forget things, and we forget selectively. We tend to forget those things which don't fit well with our established biases and preferences, and remember those things which do. We have no trouble remembering an appointment to go out for dinner and to the theater, but dental appointments easily slip the mind. In addition, time can distort the memory of observations. Since our memories are both leaky and creative, it is wise to make accurate notes when you are making your observations, a practice most good researchers have developed. Alternately, in order to avoid relying on our imperfect and creative memories, tape recordings, films, sketches, graphs, or counters may be used.

[6] James M. McCrimmon, *Writing with a Purpose* (Boston: Houghton Mifflin Company, 1967).

Selection Through Necessity

Even if our memories did not erase some observations from our minds, it is still not feasible to record completely everything that was observed. Secretaries know that exact reproduction of even a short conversation leads to a copious report. Minutes of meetings are good examples of abbreviated representations of what actually occurred. Much of the conversation must be omitted. Similarly, the researcher must select from what he remembers of the events he observed. He must decide what to record and what to omit. He would like to omit material that is not relevant to the thesis he is developing. Notice that he is not trying to discard observations that contradict his views, but to discard only those which add nothing to the point of the report. Such observations are tangential to the topic under discussion. If one wishes to describe the puberty rites of the Hopi Indians, it is perhaps irrelevant to record that homes of the Indians are constructed of thatched straw and adobe.

The researcher's bias or point of view helps him to select what is worth recording. Such biases may prevent him from considering alternatives, but are vital in terms of providing him with guidelines in selecting from the flood of data surrounding him.

Clarity and Interest

It is of no value to undertake a study and make extensive observations if the report of these observations is vague or disjointed. Since the researcher using the naturalistic observation technique often cannot rely on statistical maneuvers to summarize his data, or objective readings from a dial on some apparatus, it is crucial that his raw observations be presented clearly and concisely.

It is also a waste if a valuable bit of work remains unknown because of the pedantic or heavy-handed writing style of the authors. There is no particular reason why scientific reports should be dull, yet there is no question that many of them are so.[7]

NONPARTICIPANT OBSERVATION

The work of anthropologists such as Margaret Mead or Dorothy Eggan, who study so-called primitive societies, is based on the method of naturalistic observation. In research of this type one aspect of the method is of prime importance: nonparticipant observation. Nonparticipant observation means that the observer is a passive entity. He

---

[7] As an example of a refreshing, light style in scientific writing, see Isaac Asimov, *The Intelligent Man's Guide to Science* (New York: Basic Books, Inc., Publishers, 1960).

merely observes—he does not participate in any way in the event he is studying.

This concept applies to all methods of science, including the one under discussion. The researcher can be seen as a variable and has been studied in his own right. For example, in a well-designed and controlled laboratory experiment involving human subjects, it has been found that the sex of the experimenter can affect the results obtained. In studies dealing with inanimate matter, there is less effect of the observer on the material. It is difficult, for example, to conceive of a situation in which the sex of the geologist affects the rocks he is sampling. Still, in some experiments in physics or chemistry, the electric potential in the human body or perspiration on the hands can affect the phenomenon under study.

Smile, You're on Candid Camera

Another difficulty in research with humans, and even some other animals like monkeys, is that when individuals know they are being observed, their behavior changes. This is like the situation when we invite the boss and his wife over for dinner. We put on our company manners. Similarly, some of us know the strange effect that appearing on T.V. can have on our behavior. Even without an observable audience, the presence of the camera can throw fear into the heart of the bravest among us and stifle our spontaneity.

Agent 007

Like the well-trained spy, the experimenter or the observer tries to remain as unobtrusive as possible. He does not join in the activities of the children's play group that he is studying; he tries not to mix his own bacteria with the bacillus he is cultivating; he does not help the rat find its way out of the maze. He is a passive observer. He plays as minute a role as possible in the event he is studying. He is in essence trying to rule himself out as an event which could affect the phenomenon he is observing. For this reason, naturalists studying the behavior of birds or animals may hide themselves in blinds or use cameras, and psychologists and sociologists use the one-way mirror. When we try to study humans (with or without the aid of the one-way mirror), it is very difficult to disguise the fact that they are on display and we are observing, judging, and interpreting their behavior.

In a novel attempt to overcome this difficulty, some investigators have used confederates as undercover men who infiltrate and become participant members of the group.[8]

[8] W. F. Whyte, Jr., *Street Corner Society* (Chicago: University of Chicago Press, 1943).

Reducing Error

One procedure commonly used by researchers employing the naturalistic observation technique to counteract what we might call the "on-stage effect," is to observe the group (be present in the group), for some time before actually beginning the study. This gives the group members a chance to acclimatize themselves to the outsider. Eventually the sight of the researcher is so common that it is not an event worthy of comment. Another procedure is to have two or more researchers observing the same ceremony or event at different times. If their observations are discrepant, one explanation for the discrepancy might be that the two observers had different effects on the group.

CLASSIFICATION

The development of the periodic table in chemistry represents a most fruitful use of one aspect of the method of naturalistic observation, that of classification. In the nineteenth century, elements were being discovered rapidly but, as each element had different properties, no sense of order or relationships could be perceived among elements. Attempts to classify, group, or arrange elements into some kind of order were made by several investigators, including both chemists and a geologist. Mendeleev's periodic table was the most successful attempt at classification. Mendeleev believed that the properties of elements were more important than their atomic number; when this arrangement would not work neatly, Mendeleev left holes in the table for elements still to be discovered. He even predicted, on the basis of his table, the properties of some of the missing elements. With knowledge of the hypothesized properties of these missing elements, their eventual discovery was enhanced. Thus, the value of classification in the progress of a science.

The development of the system for the classification of plants and animals by Carolus Linnaeus (1707–1778), the Swedish botanist, is another famous example of the fruitfulness of taxonomy to science. Classifications of organisms, both past and present, into kingdoms, phylla, classes, orders, families, genera, and species is an obvious example of an attempt to replace disorder and confusion with order. To develop such a taxonomic system, close observation of the properties of organisms had to be undertaken—again the method of naturalistic observation proved invaluable. Perhaps without such a system of classification, the theory of evolution would never have occurred to Lamarck or Darwin.

The construction of classification systems is very much a function of bias or point of view. In fact, classification might be seen as a primitive form of theorizing which allows for the expression of opinion or inference. With any set of data, there are usually a variety of ways in which these data may be ordered or grouped, and the researcher's hunches or biases determine which particular grouping he will develop.

CONCLUSION

It should not be surmised that because naturalistic observation is employed to study somewhat gross behavior units, and requires little in the way of elaborate equipment, it is no longer in vogue or has outlived its usefulness, or it is a simple method and needs little training. All of these inferences are erroneous. There are many areas of study in which it is impossible, for ethical, moral, political, or practical reasons, for the researcher to manipulate events or to experiment. In addition, certain types of information can only be obtained in field or naturalistic studies. For example, if we want to investigate the phenomenon of hibernation, or famine, or juvenile gangs, we must use naturalistic observation for at least some portion of our study. The work of Piaget, the famous child psychologist, and that of Sigmund Freud was based on the method of naturalistic observation, and is the foundation for much of the current research in some areas of psychology today. Although at times this method is comparable to a crude sieve, it is still a valuable source of important and durable data in the hands of a skilled researcher.

One obvious advantage of this method is its superiority to casual observation and hit-and-miss recording or tabulating of events. A second advantage is that it does not require that events necessarily be manipulated or controlled; therefore all sorts of subject matters, normally taboo to experimental science, become open to this form of study. Another advantage of a thorough understanding of this method is in terms of the relationship of this approach to other methods. The four principles of naturalistic observation (detailed observation, accurate recording, nonparticipant observation, and grouping or classification) have application to all the other sieves of science in greater or lesser degrees. Often the data gathered by the method of naturalistic observation provides guidelines for later inquiry with more sophisticated sieves, as has been the case with Piaget's work. Finally, this method can be practiced by the young student scientist who has limited resources. In attempting to be analytical, passive, and accurate, it is often amazing what fascinating chunks of information we can produce.

chapter 5

The After–the–Fact Method

When something unexpected happens and we want to know how or why it happened, we go back in time and attempt to reconstruct the past—or at least those aspects of the past that seem to be connected with the event in question. (This method is somewhat different from naturalistic observation in that in the latter, we are investigating presently occurring events and are not necessarily interested in how such events are produced.) We face a variety of after-the-fact questions: Who or what led to the stomachache, the car breakdown, the fall of Rome, or the murder. Some of the suspects will be innocent, and some will be guilty. How do we sort guilty from innocent, or fact from fancy, after the fact?

A CASE FOR THE COURT

The job of the court is to attempt to sift fact from fancy, to sift durable evidence from perishable evidence. An event occurs and we trace back in time to find a likely suspect. For example, a crime is committed—a murder—and the investigators look out over the city, faced with an apparently impossible task. "Who did it?" Of all the thousands or, perhaps, millions of people out there, which one or ones played the key part in the crime? The prosecution must go back in time and attempt to tie the crime to a few selected people. First, they must sort suspects from nonsuspects, and then gradually separate the most likely suspects from least likely. They continue screening until a group of neutral people, the jury, can decide beyond a reasonable doubt whether the crime can be tied to a given suspect. The jury must decide how closely the crime and the suspect are tied together. Is the suspect, for example, a lone plotter and executioner, and therefore strongly tied to the crime? Or is he an unwitting accomplice, and therefore only tied to the crime by a thin line?

The problem facing the court is that of sifting fact from fancy after the event has occurred—or, as we will call it—after-the-fact. After-the-fact situations face us daily whether in the courtroom, the history class, the doctor's office, at an afternoon tea, a pub, or a garage. The problem is to select from a variety of suspects the one most strongly tied to, or leading to, the event of interest, whether it be a murder, a war, a pain in the stomach, an elopement, a lost football game, a car that won't start, or the causes of lung cancer.

The most common approach used in sifting suspects in these after-the-fact situations is sometimes known as the case method. It is a sieve with relatively big holes, and permits a large amount of fancy to slip through with the facts. Nevertheless, this kind of sieve can be a powerful aid in narrowing down the number of suspects or alternatives, particularly when the person using it is aware of its limitations. For we must be aware of the limitations of the particular method used in selecting the suspect; otherwise it is impossible to decide how much confidence the selection of a particular suspect warrants.

Consider the following methods of selecting suspects in a murder case:

Method 1: fortune teller using crystal ball ("Scarface is the guilty one")
Method 2: opinion of murdered man's wife ("Big Ear is the guilty one")
Method 3: pre-trial judgment of district attorney ("Hairy Hands is the guilty one")
Method 4: decision of jury ("Norton Noble is the guilty one")
Method 5: retrial decision of jury ("Norton Noble is the guilty one")

It is apparent that these five methods of selecting suspects deserve different degrees of confidence.

In science, too, different methods of selecting suspects warrant different degrees of confidence. While the case method, for example, warrants more confidence than casual opinion, it usually warrants less confidence than the controlled experiment. Since the case method, however, is used in many fields in which other methods are difficult or impossible to employ, we should examine its strengths and weaknesses carefully.

The After-the-Fact Method

This model is simple. An observation (O) is made. A murdered man is found, or a patient is diagnosed as depressed, or Elizabeth slaps Richard. Next, we attempt to decide what previous events (X's) led up to O, or what previous events were necessary in order for O to happen. In everyday language, we are asking what caused O.[1] There are usually several possible X's, that is, several alternative causes or events that could lead to a given observation, and the problem is to select the most probable one or ones: X_1?, X_2?, X_3?, . . . , X_n? When our murdered friend is first discovered, there are a large number of X's or suspects.

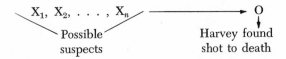

How do we narrow down the number? What aids do we use to sort the most likely suspects from the least likely ones? One aid is time. When the time of death is determined to within two hours, all suspects who have reasonable alibis for the shooting time (1:00 A.M. to 3:00 A.M.) are separated from those who do not.

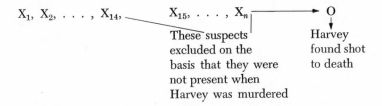

Thus perhaps all but 14 of the original suspects go free.

Let us narrow the list of suspects still further, this time using the motive aid; 12 others are now eliminated because they have no obvious motive.

X_1, X_2, X_3, . . . , X_{14}, No obvious motive \ X_{15}, . . . , X_n Clear on the basis of time that Harvey was murdered → O Harvey found shot to death

---

[1] For good reasons beyond the scope of our discussion, "cause" has become a bad word in scientific circles.

Next, Suspect 1 is focused on when it is found that the murder weapon is his. This, plus the fact that he had a motive and has no alibi, links him to the death with three lines of evidence. Nevertheless, Suspect 1 is quickly replaced by Suspect 2 when the latter's fingerprints are found on the murder weapon. As well as being the suspect most closely tied to the murder weapon, he has a motive, and has no alibi.

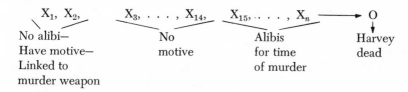

So gradually we have eliminated some X's and tied others to the crime in question—some with only one fine thread of evidence (motive) and others with several lines of evidence (motive, no alibi, murder weapon).

The case method consists of eliminating certain suspects and then attempting to determine which of the remaining ones are most heavily tied to the observation or event being studied. Its main strength probably lies in helping rule out certain suspects, rather than in guaranteeing the selection of the best suspect from among those who remain. It is not a foolproof procedure, as wrongly condemned men will hasten to tell you (at least those who still have use of their vocal cords).[2] Nevertheless, it has great advantages over the crystal ball, or the casual opinion, or the widow's bias, or the D.A.'s hunch.

A CASE FOR THE HISTORIAN

This same after-the-fact method is used by historians to explain what led to what in the past. Whereas the problem for the courts is to determine what preceding events and people are tied to a given crime, the problem for the historian is to determine what preceding events and people are tied to a given historical observation. What key events (X's) led to or were tied to the circumstances that provoked the beginning of World War II (O)? Were the key events economic, growing out of Germany's need for raw materials and markets; were they psychological, growing out of the personality structure of the German people, coupled with the resentment over their defeat in World War I; was the single key event the mistake of allowing a clever madman

[2] We noted that while none of the methods are perfect, all are designed to reduce the risk of error—the risk of certain types of error. Our courts are presumably set up to reduce the risk of condemning an innocent man on the principle that it is better to let a hundred guilty men go free than to condemn one innocent man.

to gain control during troubled times; or did the war result from a combination of all these, and if so, what relative importance should we assign to each?

$$\underbrace{X_1, X_2, \ldots, X_n}_{\text{Economic}} \quad \underbrace{1X_1, 1X_2, \ldots, 1X_n}_{\text{Psychological}} \quad \underbrace{2X_1, 2X_2, \ldots, 2X_n}_{\text{Political}} \longrightarrow \begin{array}{c} O \\ \downarrow \\ \text{Beginning} \\ \text{of World} \\ \text{War II} \end{array}$$

This is the task of the historian, to make a case for his particular selection of suspects. Because there are so many potential suspects, his job is a challenging one. This fact has led some to conclude that history is an art rather than a science. Thus any given interpretation of history must leave many suspects unnoticed, or unaccounted for; the risk of error is high enough that some decide that the conclusions of the historian might be better seen as the work of an artist rather than the work of a scientist. According to this view, the historian faced with a formidable number of suspects, develops a point of view to help him in his task. Otherwise he would literally be driven to distraction.

Many historians have been trained to look for economic suspects as explanations for historical events, and, as a result, they have missed or ignored other types of explanation. Nowadays many historians are looking for sociological or psychological suspects as well as economic ones. Thus some historians use sieves specially designed to select economic suspects; other historians use sieves designed to select sociological suspects. But if the type of suspects on the historical hit parade change, how are we to tell a good historian from a bad one? It is proposed that the good historian, like the theorist in science, is one who can sell other historians on using his theory or point of view in selecting suspects. In addition, a good historian carefully documents the suspects he has the time, inclination, and techniques to consider.[3]

A CASE FOR THE PHYSICIAN

Let us now consider another situation in which the after-the-fact or case method is employed. You have a stomach ache and you present your case to the physician. He is faced with the same problem as the court and the historian—that of narrowing down the list of suspects and then of attempting to deal with the many that remain. Just as the

[3] In an impressive effort to make their work more objective and quantitative, an increasing number of historians are adopting some of the methods of the social and behavioral sciences.

historian in his approach is governed by current theories or points of view about history, so the doctor is influenced by current medical theory, which leads him to bet on one suspect or group of suspects over others. Thus a mixture of experience, current medical theory, and personal biases determines what types of sorting devices your doctor will use, and therefore what types of suspects he will select from the many possible.

For example, what happens when you present your physician with a stomach ache?

$$X_1, X_2, X_3, X_4, X_5, X_6, \ldots, X_n \longrightarrow O$$
$$\text{Stomach ache}$$

Various alternative suspects come to his mind. Some he remembers from what he has been taught at medical school, some he has encountered often in practice, and others he remembers from reading recent articles, if he has had time to read and digest them. What are some of the suspects he might consider—appendicitis, ulcer, mild food poisoning. He does not concern himself with all of the alternatives, but initially focuses on one at a time. He examines the stomach and attempts to localize the site of discomfort. He tries to determine the time of onset, the type and degree of pain. These bits of information are applied to each of the possible common suspects. Perhaps none of the common suspects gets much support, and so all are discarded. He then commences to consider some of the possible rarer causes, using information derived from laboratory tests, X-rays, or discussions with a specialist. And so he is reducing the number of suspects and trying to select the most probable one from those that fall within the scope of his medical vision.

We have considered examples from the fields of law, history, and medicine in which a large-hole sieve—the after-the-fact method is employed. The same method is employed in economics, political science, anthropology, sociology, and business administration. The after-the-fact method is employed when it is impossible, or too difficult, or too late to experiment.

REDUCING THE RISK OF ERROR

You will have noticed that any time we can reduce the number of suspects, we increase our chances of picking the correct one or ones. The same principle holds in making a draw for a car. Which would you prefer, to have your ticket in a drum with one thousand other tickets, or with one hundred other tickets, or with ten other tickets? Any method that helps reduce the number of suspects is of as much interest to the scientist as any method of reducing the number of tickets in the drum is to the ticket holder. Let us consider some of the general ways in which we can reduce the number of suspects.

You will recall that knowledge of the time of death of our poor friend Harvey served to rule out a large number of suspects. Knowing the time at which events happen is extremely helpful in all sorts of applications of the case method. The information provided by the time sort allows us to rule out certain suspects whether the problem is one for the court, for the historian, for the physician, or for the garage mechanic.

Consider the case of Elvis—a tall, blond, good-looking young man of 20 who is referred to a psychiatrist. Elvis had difficulty sleeping, was tense and anxious, and seemed to spend a lot of time worrying about his courses at the University, his looks, and his dates. We start then with Anxious Elvis as our observation (O).

The psychiatrist has the job of attempting to see what led to O. How many suspects are there? The number of suspected reasons or events contributing to Elvis's unhappy state are infinite. Everything that has happened from the moment of his conception is a potential suspect.

Time Scale	$X_1, X_2, X_3, \ldots, X_{40}, \ldots, X_{85}, \ldots, X_{100}, \ldots, X_{105}, \ldots, X_{110}, \ldots, X_n$	→ O Anxious Elvis now

Genetic heritage — Birth trauma — Weaning — School — Fall from bike — Failed Grade III — Strict father — Bright sister — Etc. Etc.

Elvis's problem, then, may be due to a lot of things: a genetic factor—bad genes, "his grandfather was the same way"; or his mother did not want him, handled him roughly, and weaned him too early; or a witch of a Grade I teacher who terrorized the boy; or a bang on the head received from a fall off a bike at the age of eight; or a stern father who was always smiling but never satisfied, no matter how hard Elvis tried to please; or any one of a thousand things or combination of things. Like the historian, the psychiatrist must have some biases, or theories about what events lead to anxious people. Otherwise, the avalanche of possible suspects could drive him . . . well, to see a psychiatrist. Thus some psychiatrists look to your genes and your biochemistry (using genetic or biological sorts), others to your weaning and toilet-training (using early experience sorts), others to your attitude toward authority.

But how can knowing the time at which events occur (using the time sort) help reduce the overwhelming task facing the psychiatrist? If Elvis has been anxious for years it is difficult to know what factors are important. If, however, he has only been anxious for a month, think how this reduction in time narrows down the number of suspects!

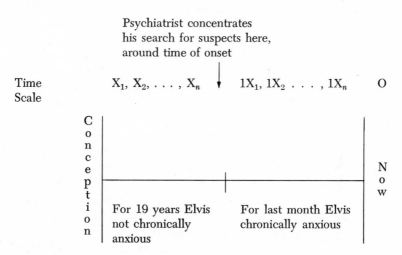

Psychiatrist concentrates
his search for suspects here,
around time of onset

Time Scale $X_1, X_2, \ldots, X_n$ $1X_1, 1X_2 \ldots, 1X_n$ O

Conception

For 19 years Elvis not chronically anxious   For last month Elvis chronically anxious

Now

Now the psychiatrist has a better chance to home in on some contributing suspects by examining events just preceding the onset of anxiety. This is not to say that events during the first 19 years are irrelevant, but at least they were not sufficient alone to lead to an anxiety state.

SUMMARY

Before we proceed to an extension of the case method, a summary is in order. The case method is an attempt to go back in time after an event has occurred, and find the most probable cause or suspect. There are at least two major problems: (1) there are usually a great number of possible causes or suspects; (2) it is often extremely difficult to get reliable information about past events or symptoms (possible causes or suspects). For example, the courts must rely heavily on the memory of the witnesses; and such memories have been proven time and again to be unreliable. Also, much of the evidence is unavailable due to eye-witnesses melting into the crowd, or failing to volunteer evidence for fear of getting involved in lengthy or unpleasant court proceedings. Your doctor, too, relies heavily on the fragmentary bits of information you give him—bits you select from the large file of fact and fancy which you call your memory, that wonderful data swamp. The historian also deals in fragments of the past, documents perhaps written from the fallible recollections of someone who knew someone who saw "it" happen. We are rarely sure how time has taken an event and shaped and reshaped it to fit the minds and tongues through which it passes. If you have ever played the party game, "Rumor," in which a message is passed from person to person in the same room, you know how quickly the message can be unwittingly distorted and reshaped by each person in the chain.

Thus the case method, with its large numbers of suspects—some lost to view, some distorted by time—presents a great challenge to those who would sort fact from fancy. With an awareness of the difficulties, however, and in the knowledge that no more practical method is readily available, dedicated men in law, medicine, and history have demonstrated that useful and durable packages of information can be produced. Other men in the same fields, men less dedicated or less aware of the limitations of the method, have posed and have been accepted as authorities, only to be unmasked later as pushers of perishable goods. Such people are a double threat: a threat to themselves, in that they assume they know and so are less open to other evidence, and a threat to those who rely on their advice. Perhaps the best safeguard exists when both the experts and the public are aware of the strengths and weaknesses of the various methods used to sift fact from fancy.

In spite of its limitations, do not reject this after-the-fact method of attempting to sift fact from fancy simply because it cannot guarantee to produce the correct suspect. In many situations it remains the best method we have. The method is useful in narrowing down the field of suspects so that we can at least make a decision—probably a more durable one than those produced by the rules of evidence of the armchair, the pub, or the afternoon tea party.

In the next chapter, we will consider an extension of the case method. It can be called the before-and-after method, and can be applied in situations that do not involve going back into the past for suspects.

chapter 6

Before–and–
After Method

With the method of naturalistic observation, discussed in Chapter 4, the aim of the researcher is to focus, with a wide-angle lens, on raw data in its natural surroundings, rather than in the library, examining room, or laboratory. The researcher's aim is careful semantic description, whether he is studying the mating behavior of the stickleback or the fighting behavior of street corner gangs. The researcher in this instance is not primarily concerned with "why" or "how come," although he may suggest theoretical explanations. In contrast, the after-the-fact sieve, considered in the last chapter, focuses on the past rather than the here and now—still using a wide-angle lens, however. The method deals essentially with samples of pre-packaged data secured from memories, written records, and other bits and pieces that can be dredged up. Also, with the after-the-fact method, the researcher is frequently concerned with questions of "why," whether they pertain to World War II, or an unexpected death.

The before-and-after method applies to the type of questions which start out with a statement such as "I wonder what will happen if . . . ." That is, you start with O_1, the onset of a rash of pimples, and then introduce X, a new skin lotion—Hornet Honey, and after the prescribed three-month treatment you make another observation O_2, and report complexion improved. The question is, "Did the Hornet Honey help?" or was it one or more of the other X's in the stream of events taking place between O_1 and O_2 that did the trick?

$$O_1 \qquad X_1, X_2, X_3, X_4, \ldots, X_{27}, \ldots, X_n \longrightarrow O_2$$

O_1	Hornet	Water	Improved
Onset	Honey	softener	complexion
of		installed	
pimples			

Can you think of some other suspects? In that given period there are many other suspected reasons for the rash clearing up: the person finished exams, or started holidays, or got a new boyfriend, or the family installed a water softener, or the person stopped eating so much junk between meals, or finished some biochemical growing up, or got more sunshine, or got used to the rash so that even a small improvement looked great. But "Hornet Honey" will be only too happy to take credit for the improvement, and probably, in such a situation, most of us would be quite content to give the salve a testimonial, even though as we have just indicated it is only one of many suspects. Many medical treatments that are given credit for cures are likewise one of many suspects, and it requires a comprehensive series of experiments, of the kind to be discussed shortly, to obtain durable information about the adequacy of the supposed treatment.

This before-and-after method we have just considered would not be regarded as scientific by some researchers because too many suspects always remain even after we have ruled out many others. But suspects always remain, so whether a method is or is not scientific is a relative question. Four main types of suspects, or rogues, challenge us in our attempts to sort fact from fancy, in our attempts to locate or manufacture stable packages of information.[1]

The following medical example will allow us to identify these four troublemakers. After exposing them, we will consider various tactics we can use to minimize the mischief they can do.

---

[1] Donald T. Campbell and Julian C. Stanley, *Experimental and Quasi-Experimental Designs for Research* (Chicago: Rand McNally & Co., 1967).

FOUR MISLEADING SUSPECTS

A doctor has been given the job of evaluating a new drug for treating depressed patients in a hospital. He examines the patients before the treatment (O_1) and after the treatment (O_2). The treatment (X_d) consists of a six-week period during which two of the new pills were taken three times a day. After the treatment period the doctor decides that most of the patients—75 percent of them—have improved.

$$O_1 \qquad X_1, \ldots, X_2, \ldots, X_d, \ldots, X_7, \ldots, X_n \qquad O_2$$

All patients	New drug	75% of
in group	treatment	patients
are depressed		improved

The doctor is very pleased and writes an article for a medical journal. A medical researcher reads the article and writes to the doctor, saying there are at least four other explanations as to why the patients improved, reasons which the doctor apparently did not consider.

In-the-gap Suspects

Were there any changes in ward routine, or ward staff introduced during the treatment period that may have contributed to patient welfare? A new cook or a new ward supervisor may have been more influential than the drug in bringing about patient improvement. It is important to remember that after careful study many wonder drugs turn out to be duds. Since the doctor was looking at the drug, and not at other possibilities, the drug was given credit and the other possible suspects went unnoticed.

In other words, many events which took place between O_1 and O_2, in addition to the drug, are legitimate suspects in contributing to the cure. We will call those types of suspects "in-the-gap" suspects.

Time-tied Suspects

The researcher asks another question. Would 75 percent of the patients have improved in a six-week period even without treatment? Some illnesses cure themselves given time alone—that is, variables such as natural recovery time, in the case we just mentioned, or maturity, in the case of complexion changes during adolescence. In many instances of the simple before-and-after design, time alone may deserve the credit for the improvement rather than the pet treatment of a given investigator. Thus any suspects that are specifically related to the time between O_1 and O_2 have been called "time-tied" variables or suspects.

Elastic Ruler Suspects

The researcher mentions a third explanation apart from the drug: namely, the doctor's ability to measure depression may have changed. We cannot measure depression with a precise ruler, rather we measure it with a man's judgment which, as we all know, fluctuates from time to time like an elastic ruler. If, for example, the doctor wanted the drug to work, this could affect his judgments so that he actually would imagine improvement whereas an unbiased doctor would not. Those things that influence the measuring instrument or the person doing the measuring we call "elastic ruler" variables or suspects.

On-stage Suspects

Finally, the researcher suggests yet another factor (other than the drug) which might explain the patients' improvement. The very act of interviewing the patients at the beginning to see how sick they were may have influenced them, particularly if they knew what was happening. Some people respond for a while to almost any new treatment. On the second interview they might well want to appear better so as to be able to go home, or to help the nice young doctor. Thus, in some cases when people know they are being watched or measured, their behavior changes; they are not themselves. This type of effect we call "on-stage" variables or suspects.

These four types of suspects plague every before-and-after study. Unless we can separate out their effects from the effects of the treatment we want to study, we can seldom be sure whether the treatment is good or not. The control group method goes a long way toward bringing these four rogue suspects under control. This method will be examined in the next chapter.

Table 2. COMMON SUSPECT TYPES

TYPE	EXAMPLES
In-the-gap	Exam
	Stock market crash
	Catching cold
	Mother dies
	Hangover

Table 2. COMMON SUSPECT TYPES *(Cont.)*

TYPE	EXAMPLES
Time-tied	Hungry Tired Older Rested Menstrual period
Elastic ruler	Boredom, fatigue, or mood of researcher Instrument wear or breakdown Bias of researcher, practice
On-stage	Recall Putting best foot forward Lying

SKILLFUL USE OF THE BEFORE-AND-AFTER SIEVE

The before-and-after model, such as the case of the onset of the rash of pimples, can be used to advantage when: (1) the observation under study has remained stable, or not elastic, for a period of time; (2) a quick-acting remedy is being tested; and (3) the number of suspects pouring into the gap between O_1 and O_2 can be carefully controlled. Let us consider the first instance, the one in which the observation remains the same for an appreciable period of time.

Stable Observations

This can be portrayed as follows.

$$X_1, X_2, \ldots, X_n \qquad 1X_1, 1X_2, \ldots, 1X_n \qquad 2X_1, 2X_2, \ldots, 2X_n \qquad 3X_1, 3X_2, \ldots, 3X_n$$

Hornet Honey

$$O_1 \;=\; O_2 \;=\; O_3 \;=\; O_4 \;\neq\; O_5$$

Rash   Rash   Rash   Rash   No rash

In this instance O_1 equals O_2 equals O_3 equals O_4. Then after the fourth observation, we try the new treatment, the Hornet Honey, and find a marked improvement in the rash of pimples. That is, O_4 does not equal O_5. Now we probably have more justification in getting excited about Hornet Honey because the rash has been exposed to a wide variety of other suspects between O_1 and O_4 with no obvious change. Similarly, the other main suspects (in-the-gap, time-tied, and on-stage effects) have had considerable opportunity to work prior to O_5 without any evidence that they are important suspects in this instance. The possibility still remains that there would have been a change in O_5 even if Hornet Honey had not been used—that is, that the natural course of the rash of pimples had run out (a long time-tied suspect at work)—and that what we are seeing is a spontaneous recovery that happened to coincide with or follow the administration of the Hornet Honey. Let us suppose, however, that following O_5 we stop the treatment with the Hornet Honey and that at O_6 the pimples have returned again. Now we reintroduce Hornet Honey, and at O_7 the pimples have gone away. At this point we have increased confidence that the change is somehow connected to the taking of Hornet Honey. It may well be that the person's anxiety has been reduced, so that had he taken anything in which he had confidence the same results may have been obtained, or it may have been directly related to some chemical in the Honey. Nevertheless, we have confidence that some part of our treatment ritual is useful. You can probably figure out how to continue the study in order to have increased confidence that the improvement depends specifically on Hornet Honey.

Quick-Acting Suspects

It should also be noted that the before-and-after method has considerable power when we feed a quick-acting and powerful suspect or X into the gap between O_1 and O_2. For example, it is much easier to detect a cure for depression that takes one day to work than one that takes three months because there are fewer alternative suspects to pour into the 24-hour gap than the three-month gap (n is much smaller than n^1).[2]

O_1 $X_1, X_2, \ldots , X_d, \ldots , X_n$ O_2

$\longleftarrow$ ———————— 1 day ———————— $\longrightarrow$

O_1 $X_1, X_2, \ldots , X_d, \ldots \ldots \ldots \ldots \ldots \ldots \ldots \ldots \ldots , X_n^1$ O_2

$\longleftarrow$ ———————————— 3 months ———————————— $\longrightarrow$

Penicillin is an example of a quick-acting X, the effects of which are relatively easy to detect. That is, with the administration of such an antibiotic the temperature quickly returns to normal, or the infection very soon shows obvious signs of improvement. Some of the unpleasant side effects, which appeared more slowly, however, took longer to tie to the antibiotic—again, the longer the time interval between X and its effect, the more difficult the research problem. Think, for example, how long

---

[2] There is also less time for time-tied and elastic ruler factors to operate.

it must have taken primitive man to understand the relationship between the birth of a child and the act of procreation.

Thus the slow-acting X's, whether diseases or treatments, are usually the most difficult to detect. If the before-and-after method is to be used in such instances, we must be prepared for relatively slow progress and a period of trial-and-error treatment with favorite, but often fictional, cures.

Keeping Stray Suspects Out

Obviously if we can greatly reduce the number of stray suspects pouring into the gap between O_1 and O_2, we simplify our task. In economics, political science, sociology, medicine, and psychology, it is often difficult to control the flow of stray suspects into the gap. The method gains in power, however, in those investigations in which you can carefully shield the gap between O_1 and O_2 from almost all suspects except the one you wish to study. Furthermore, if you can control for elastic ruler effects, on-stage effects, and time-tied effects, you have a very powerful investigative method.

Certain observations in physics and chemistry meet these conditions. Let's examine a hypothetical example. A physicist may wish to test the effects of heating a compound on its radioactivity.[3] He places a Geiger counter near the compound, takes the reading on the Geiger counter, then heats the compound to the desired temperature, takes another reading, and notices that the radiation has decreased.

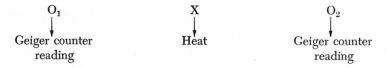

Now you ask the physicist how he knows it was the heat, and not a reduction in stray sources of radiation from the sun, or a wrist watch dial, or dust from bomb-testing fallout, that resulted in the reduced readings, since the Geiger counter was exposed to all of these, as well as to the compound under test. Or you suggest that the change in reading is due to an elastic ruler effect—the batteries in the Geiger counter wearing out between O_1 and O_2.

If he was a careful researcher, he would say that he had placed the compound and the counter in a shielded box so that only the rays from the compound being tested could operate on the counter. He would indicate that he had tested the sensitivity of the counter before and after

[3] Actually heat has very little effect on the radioactivity of a compound, so it requires a carefully controlled experiment to measure such small effects.

the study and found no change, thus indicating there had been no elastic ruler effects of any consequence. Also, there is little concern over time-tied suspects because the natural reduction in radiation of the compound under study is very slow with appreciable changes taking years, not minutes. Finally, there is no evidence that on-stage effects are operating, that is, that the measuring process itself affects the compound in any way—it does not radiate less at the end of the study just to please the nice young physicist! A critic suggests that the effect was not due to a physical change, but to a chemical change—the heat led to the giving off of radioactive gas which escaped from the chamber in which the counting was done. However, the student can test this by redesigning the chamber.

While it is true that many studies in the physical sciences require extensive and precise equipment, relatively simple research methods, such as the before-and-after method, are adequate for two reasons even though this method is more difficult to use in the behavioral sciences. First, as we noted, it is usually easier to control the relevant rogue suspects in physical science. Second, the rogue suspects are usually fewer and better known. Nevertheless, as the physical sciences deal increasingly with less stable materials, researchers must use more complex research methods and statistical procedures.

Alternatively, some behavioral scientists have attempted to add precision to their work by restricting their studies to more or less stable materials, like white rats, under laboratory conditions where factors like diet, weight, love life, and the cost of living can either be controlled, or assumed to be irrelevant to the rat. In behavioral science situations where the rogue suspects cannot be directly controlled, or assumed to be irrelevant, we often require more complex designs, such as the control group method, which the next chapter discusses.

chapter 7

The Control Group Method

Earlier in the discussion it was indicated that there was a method which represented a major breakthrough in helping us deal with the four rogues: in-the-gap, time-tied, elastic ruler, and on-stage. Consider what would happen if we use two groups of patients instead of one when evaluating a new drug treatment.

We split the group of depressed patients in half, making two groups. One group gets the new wonder pill and the other group gets a pill that looks exactly the same but contains only sugar. Some people feel better if they take a pill—any pill—so we must see whether more people get better in the wonder drug group than in the sugar pill group. We divide the group in such a way that there is little chance of getting healthier patients into one group than in the other. Ideally, the groups should start out identical in as many respects as possible. One way to protect yourself from bias in your selection is to pick the names out of a hat, the first name goes to Group 1, the second name to Group 2, the third to Group

1, and so on. Unless you have some way of protecting yourself against your own personal bias or the biases of others in the institution, you end up with a special collection of patients in one group (e.g., staff members attempting to get their patients or relatives into the group that will get the new pill).

Once we have two groups which are comparable as far as we can tell, we treat them exactly the same way with one exception: one group gets the wonder pill, while the other group gets the sugar pill or placebo.[1]

The procedure now becomes:

Group 1 O_1 $X_1, X_2, \ldots, X_d, \ldots \ldots \ldots \ldots, X_n$ O_2
                                          New Drug

Group 2 O_{1a} $X_1, X_2, \ldots, X_p, \ldots \ldots \ldots \ldots, X_n$ O_{2a}
                                          Sugar Pill

Thus the two groups start out supposedly with the same amount of depression, that is, $O_1 = O_{1a}$ or, in other words, one group does not have more seriously depressed patients than the other.

EQUAL SUSPECT EFFECTS

In-the-gap

The main point is to try to run the experiment so that the individuals in the two groups are treated exactly the same except for one suspect, the drug X_d. Thus, we attempt to make sure that the same X's pour in-the-gap between O_1 and O_2 for both groups—the same cook, the same ward supervisor, the same ward, the same weather, the same amount of time between O_1 and O_2 as between O_{1a} and O_{2a}. To ensure that the nursing staff does not spend more time with the patients in one group than with the patients in the other, the patients in the two groups are mixed up or made indistinguishable as far as anyone who can influence the experiment is concerned. The nurses and other doctors are not told which patients are getting the new wonder drug, X_d, and which ones are getting the sugar pill, X_p. All patients receive a pill that looks identical.

Elastic Ruler

If the experiment is run properly, the doctor who measured the depression at the beginning and at the end does not know which

---

[1] In some studies instead of giving a sugar pill, the researcher uses the treatment pill in common use. Thus he can see whether more people improve in the new pill group than in the group receiving the usual treatment.

patients have received X_d and which got X_p, so his biases, or elastic ruler, cannot influence his assessment of one group over the other either during the study or when he is deciding which patients have improved and which have not.

On-Stage

The on-stage effects of having been interviewed would influence the patients in both groups. There will be patients, no doubt, in both groups who want to impress the doctor that they are well enough to go home, as well as some who want to help the nice young doctor. Thus we hope that the resulting influence on O_2 and O_{2a} will be about the same; that is, that both groups will probably show some on-stage improvement apart from any effect of the new drug.

Time-tied

Furthermore, spontaneous recovery should be about the same for both groups, since the time between O_1 and O_2 is the same as between O_{1a} and O_{2a}. Thus time-tied or natural recovery suspects should affect each group the same way.

NOW WHAT

We should not be surprised if both groups show some improvement; O_2 shows an improvement over O_1 and O_{2a} shows an improvement over O_{1a}. These changes reflect the effects of such suspects as spontaneous recovery (time-tied), biased doctor (elastic ruler), nice ward supervisor (in-the-gap), desire to go home or to help the nice young doctor (on-stage). But if X_d has had an effect greater than X_p, we should have O_2 showing a greater shift than O_{2a}. If the two groups were the same to begin with, the difference between O_2 and O_{2a} provides us with a measure of the effect of X_d over X_p. Whereas with the simple before-and-after model:

$$O_1 \quad X_1, \ldots, X_2, \ldots, X_d, \ldots \ldots \ldots, X_n \quad O_2$$

we have X_d effects all mixed up with the effects of the other suspects, and it is difficult to untangle them. It was this kind of tangle that led some wit to wisely observe that a good doctor is one who keeps the patient occupied while nature works the cure. It is easier to wait for a natural change when one is given the impression that something is being done to bring it about.

When we divide a group in two in a way to make the two sections as identical as possible, and then give them the same treatment except for one X, we are using a *control group* method or procedure. It is a more precise sieve than the naturalistic observation, the after-the-fact and before-and-after sieves so far discussed. It represents a remarkable leap forward in helping us produce packages of durable information. In one stroke it permits the researcher to assess the effects of his treatment over and above the effects of those four great rogues: in-the-gap suspects, time-tied suspects, elastic ruler suspects, and on-stage suspects.

ELABORATION OF CONTROL GROUP SIEVE[2]

Suppose that, in the example just discussed, all depressed patients had been given a stimulant pill each day as part of the regular hospital routine, and thus at the time of the study Group 1 individuals were given both the new wonder drug as well as their regular pep-me-up pill. Group 2 patients, however, were administered the stimulant drug only.

Group 1 X_s O_1 $X_1, X_2, \ldots, X_d, \ldots, X_n$ O_2
      Stimulant   Wonder Drug

Group 2 X_s O_{1a} $X_1, X_2, \ldots, X_p, \ldots, X_n$ O_{2a}
      Stimulant   Sugar Pill

Can we conclude that if O_2 is different from O_{2a} the difference is due to the wonder drug? One is tempted to answer yes and to argue that, since both groups were given the stimulant, any difference between the groups must be due to the difference between X_d and X_p. It is possible, however, that it was the mixture, combination, or interaction of the stimulant and the wonder drug that led to improvement, and that the wonder drug alone may be ineffective. In this instance we could perhaps repeat the study with two other groups of patients and arrange that no stimulant pills be given and thereby determine what improvement would result from the wonder drug alone. Sometimes, however, the effect of making the first observation may get mixed or combined with the treatment, and we may want to know what the effect of observing or measuring alone has—as well as the treatment effects alone—as well as the interaction of the two.[3]

---

[2] Campbell and Stanley, *Experimental and Quasi-Experimental Designs.*

[3] The distinction between O and X can become somewhat vague or arbitrary. In essence, an O may be considered as a suspect, or X, when the first O has effects on subsequent observations.

Consider another example. Assume we are interested in determining whether providing children with training in physical coordination will improve their intellectual ability. We could design a control group study to test this. In order to ensure that our two groups are equal in intellectual ability to begin with, we administer an intelligence test to all the children and divide them into two groups with similar numbers of bright, average, and dull children in each group. Group 1 is then given two days of training and subsequently both groups are retested.

Group 1	O_1	$X_1, X_2, \ldots, X_t, \ldots, X_n$	O_2
	Intelligence test	Training in coordination	Intelligence test
Group 2	O_{1a}	$X_1, X_2, \ldots \ldots \ldots, X_n$	O_{2a}
	Intelligence test		Intelligence test

If the difference between O_2 and O_1 is greater than the difference between O_{2a} and O_{1a}, we may conclude that training in physical skills improves intelligence test scores. But a sophisticated critic argues, "You may have controlled for the four rogues, but you still don't know whether the improvement is due to a combination or interaction of the pretest (O_1) and the training (X_t). Maybe the children in the first group were just more familiar with the experimental setting. The relaxation that comes with familiarity, plus the extra attention during training, may have produced the difference between O_2 and O_{2a}."

How can we answer this critic? One way is to randomly select two extra groups of children from the same classroom. Neither group is given the first intelligence test, but we assume, since they were picked from the same classroom, that they would have the same average scores as the other groups. In addition, we give training to one of the groups, but not the other. The model thus becomes:

Group 1	O_1 Test	$X_1, X_2, \ldots, X_t, \ldots, X_n$ Training	O_2 Test
Group 2	O_{1a} Test	$X_1, X_2, \ldots \ldots \ldots, X_n$	O_{2a} Test
Group 3	Randomly Assigned	$X_1, X_2, \ldots, X_t, \ldots, X_n$ Training	O_{1b}
Group 4	Randomly Assigned	$X_1, X_2, \ldots \ldots \ldots, X_n$	O_{1c}

Assume that we obtain the following results:

$$O_1 = 60 \qquad\qquad O_2 = 110 \qquad\qquad O_{1b} = 85$$

$$O_{1a} = 60 \qquad\qquad O_{2a} = 80 \qquad\qquad O_{1c} = 60$$

If we compare O_2 with O_{2a}, we have a measure of the effectiveness of training combined with a pretest—30 units difference. A comparison of O_{1b} with O_{1c} gives an indication of the effect of the treatment alone—25 units difference. The difference between O_{1c} and O_{2a} gives a measure of the pretest effect—20 units difference.

We now conclude that training-plus-test practice accounted for the greatest improvement. Testing alone and training alone were not as effective as the two combined. Knowing the size of the test effect alone and the training effect alone would not lead to the correct prediction regarding the size of the two effects together.

When there is a strong possibility of an interaction between X and O, it is advisable to test the effects of each separately.

OVERVIEW

We are now in a position to make several important points. Naturalistic observation is a starting point for science. Distinguished from casual observation, naturalistic observation introduces the following data collection, storing, and sorting tools: (1) detailed and concentrated observation, (2) accurate recording and record keeping, (3) non-participant observation, and (4) identification of patterns or classes among the carefully recorded, detailed observations. All of our scientific methods rely on these four basic tools.

Even with these tools, however, the four rogue suspects (in-the-gap, time-tied, elastic ruler, and on-stage) play a part in the data-packaging operation. The after-the-fact method, and the before-and-after method can be seen as forms of naturalistic observation, or extensions of it, in which the confounding effects of the four rogue suspects clearly emerge. In those instances where the four rogue suspects can be directly controlled, such as some experiments in physics and chemistry, the after-the-fact method and the before-and-after method have considerable power. The before-and-after model also has impressive power under conditions involving a quick-acting and strong suspect and/or under conditions when O has remained stable for some time in spite of attempts to modify it. Otherwise the before-and-after method is relatively less efficient, and whenever possible the control group method should be used. In this method the four rogue suspects, while present, should affect the treatment and nontreatment groups in approximately the same way. Thus, if the treatment has an appreciable effect over and above the effects of the four rogues, you have a chance of demonstrating it. However, as we noted, there are instances in which the control group method is not applicable, and in which the case method or the before-and-after method are our only methods of making decisions (other than relying on the rules of evidence of the armchair, the pub, busy committees, or imperfect memory combined with personal bias). If the case method or the before-and-after method are to be used, it is important that both the people who apply them, and also the people who base their decisions on the findings, should be aware that there is likely to be a good mix of bias and chunks of perishable data included in the findings. With such awareness we are in a position to decide how much confidence we are prepared to place in the packages of information these methods produce.

Summarizing, then, we can say: (1) Science has no perfect methods for collecting and packaging knowledge or information; (2) scientists use a variety of methods which differ in precision and cost; (3) a sieve or a scientific method is useful to the extent that it assists a scientist to make decisions by reducing the number of suspects involved; (4) it is not always possible or wise to use a fine sieve on a problem—a coarse sieve often becomes acceptable if it reduces ignorance even a little bit.

I. Naturalistic Observation

$$O_1, O_2, O_3, O_4, O_5, O_6, \ldots , O_n$$

II. After-the-Fact Method

$$X_1, X_2, \ldots \ldots \ldots \ldots \ldots \ldots , X_n \qquad O_1$$

Number of possible suspects large and often difficult to identify or locate

III. Before-and-After Method

$$O_1 \ldots 1X_1, 1X_2, \ldots , X_t, \ldots , 1X_n \qquad O_2$$

Reduces number of suspects but still can't untangle in-the-gap, time-tied, elastic ruler and on-stage suspects from treatment or X_t under study

IV. Control Group Method

Group 1 $\quad O_1 \ldots 1X_1, 1X_2, \ldots , X_t, \ldots , 1X_n, \ldots \qquad O_2$

Group 2 $\quad O_{1a} \ldots 1X_1, 1X_2, \ldots \ldots \ldots , 1X_n, \ldots \qquad O_{2a}$

Reduced number of suspects. Major effects of the four rogue suspects should influence Group 1 and Group 2 about the same. So if groups start out the same and end up different, we assume the difference is due to X_t.

CURRENT USE OF FINE HOLED SIEVES

We have attempted to demonstrate in this section the plays used by scientists in their game of sifting durable from perishable data. We have seen that some of the plays, although crude, still score points and we have learned other plays which, although not guaranteeing a major score, reduce the chances of a big loss enormously.

To what extent do researchers in the behavioral sciences use the high cost elaboration of the control group model—the one most likely to provide facts unclouded by the mist of fiction? Rarely is this method used. The most popular method of research is to assign people to groups and assume that, because the individuals were selected from the same area, school, or institution, they will be equal in severity of depression or intelligence or whatever it is the researcher is investigating. Or, alternately, we select our groups because we know they differ in some dimension and we may want to know how this difference affects certain behaviors. Thus the common model is

| Group 1 | Rich children | $X_1, X_2, \ldots, X_t$ | O_1 |
| Group 2 | Poor children | $X_1, X_2, \ldots\ldots$ | O_{1a} |

Why don't we use the more elaborate model? The answer is we are not prepared to pay the cost. As we move along the continuum from the large hole to the small hole sieves, the resources required for effective use of the method increase, too. The scientist using the naturalistic observation or after-the-fact method can often work alone, requiring little in the way of fancy space or equipment. Certainly there are applications of each of these methods where phenomenal resources are used, but it is possible to do a study with minimal computing costs. With the other methods we are faced with questions of coordination, administration, and research time that (in the case of the elaboration of the control group method particularly) boggle the mind.

We propose that the researcher selects the method of data collection and analysis that (1) is acceptable to the science game referees (journal editors, thesis committee members), (2) will likely provide him with data that is relevant to his interests, (3) will likely provide him with some durable data, and (4) can be used with the time, energy, and resources available.

Therefore, it is quite likely that the researcher often uses other methods than he would prefer simply because of limited resources—equipment, manpower, time, etc. It is equally likely, however, that, because of habit, he uses methods that are often inappropriate. Like all of us, researchers develop comfortable rituals, data collection rituals and data analysis rituals, that are effortless and so are used whether they are particularly appropriate or not—like the man who uses a sledgehammer both for smashing boulders and for cracking nuts.

three

The
Magic of
Numbers

LANGUAGE, we have seen, is a set of symbols, sounds, or written squiggles, and rules for combining these symbols. In English, the symbols are words and the rules are our grammer.

Measurement is a language, too; its symbols are numerals. In this section we will be talking about the different ways in which numerals are tied to objects and events to help us describe and order our world. Numbers are one of the most common methods of packaging data into memorable chunks.

Not only do we use numerals or numbers to label the sweaters of football players and to count poker chips, we also use them to tell us how we measure up in height, school grades, and income. We use numbers to help us describe things we can't see directly, like temperature, anxiety, and the national debt. Furthermore, we use numbers to help us move from small bits or samples of information to generalizations, like predicting the number of cases of cancer in the whole population on the basis of the number we find in a sample of the population.

In the following chapters, we will have an opportunity to become familiar with some of the major rules of the number game, and also to learn how we can be sadly misled when people, wittingly or unwittingly, break some of the rules.

chapter 8

The Number Game

Almost everyone knows that measurement and numbers belong together. In fact, measurement can be defined as tying numbers to objects and events according to certain rules in the same sense that words are tied to objects and events.

Measurement, unlike mathematics, is really part of the semantic system. In the case of measurement, instead of tying word labels to shareable events or objects, we tie number labels to them. Measurement also includes syntactical rules for combining numbers which are translated into semantic operations. Thus, for example, the syntactical symbol + (plus) can be semantically translated into pouring water from one beaker and adding it to the whiskey in another beaker.

NUMBER GHOSTS

In an earlier chapter we noted that different people use different rules for tying words to objects and events. At one extreme both the word ("gloobic") and the events (individual images and sensations) are highly personal in nature. At the other extreme we know that some words (sun) and the events or objects they represent are highly public in nature. A crude eight-point scale, stretching from individual pragmatics to large group syntax, was presented as a guide in deciding where on the private-public continuum different words and the objects they represent lay. Numerals are nothing but number words which we tie to objects, events, or data samples, according to rules which again may be highly personal (70° sad), or public (70° Fahrenheit), or which fall at various points between on the personal-public dimension. Since we all learn number words early in our lives, however, and since measurement carries with it an aura of precision and accuracy, there is great danger in assuming that whenever people tie number words to objects or events, they must know what they are talking about.

When objects are assigned numbers, we are less likely to attempt to find out whether the investigator is playing by pragmatic or by semantic rules. We are less likely to apply the acid test, "can different investigators independently assign the same numbers to the same objects and events?" We are less likely to remember that a ghost remains a ghost whether he is called "Casper" or "G97831.42."

The Dominion Bureau of Statistics in Canada keeps records of the diagnoses of hospital patients. A different number code is used for each diagnosis. For example, any patient with epilepsy may be assigned the number "353," any patient with polio may be labelled with the number "344," and the patient with schizophrenia may be assigned the number "300." Now we know that doctors often disagree about the diagnosis tied to a given patient. Such disagreements suggest they are playing by pragmatic or personal rules. The rules are no less personal when the doctor assigns the number "300" to the patient than when he assigns the word "schizophrenia."

In making up the reports for the Bureau of Statistics, the job of translating the diagnosis into the number code is often left to stenographers and clerks. The stenographer goes through the patient's file, locates the diagnosis, then finds the appropriate number code in the Bureau of Statistics Handbook, and enters that number code as the patient's diagnosis. However, since doctors do disagree about diagnoses, and since patients often see several doctors, the stenographer is faced with the problem that there are frequently several diagnoses in a given patient's file. We would assume that in such an instance she would consult a physician to find out what the real diagnosis is. If she does, she is often merely getting his personal preference for the name to be assigned to this diagnostic ghost—and this is what the Dominion Bureau of Statistics will get—a numbered ghost. But the stenographer in one particular hospital had been taught to use her own judgment since, like most of us, physicians are only too happy to delegate responsibility when they're not clear about what to do themselves. Our stenographer finds that a patient has two diagnoses, one of which can be assigned the code number "308" (suspected brain tumor) and the other which can be assigned the code number "300" (schizophrenia). Being most ingenious, she simply adds the two numbers together (300 + 308 = 608), takes the average by dividing by two and assigns the patient the code number "304" which happens to stand for senile. Thus

$$\frac{1 \text{ schizophrenia} + 1 \text{ tumor}}{2} = 1 \text{ senile}$$

With this approach, we have a 24 year old male patient who has been diagnosed as "brain tumor" (308) by one psychiatrist and "schizophrenic" (300) by another psychiatrist and who ends up with the diagnosis "senile," or old age (304), an ailment not frequently found in 24 year olds.

The point is that whether words or numerals are used, the Dominion Bureau of Statistics ends up with a massive file of diagnostic ghosts. Researchers rely on these files to write scholarly reports. They attend international meetings at which they solemnly compare different provinces and different countries in terms of the number of schizophrenics, the number of patients with brain tumors and the number diagnosed as senile. They develop elaborate theories as to why there are more schizophrenics in one region than another. It is fascinating to speculate that these researchers may, in many instances, be unwittingly studying the habits of hospital stenographers rather than the incidence of various illnesses.

Thus, whenever you read a piece of research, it is to your advantage to ask the question, "What evidence is there that the researcher is not just counting ghosts?" You are asking, in essence, what evidence does the researcher present to indicate that independent observers could do the original counting, or measuring, and come up with the same numbers. If the researcher does not present such evidence, you will be left wondering (a) whether the researcher has been spending his time playing a personal game of ghost counting, or (b) whether the researcher has in fact been measuring or counting shareable objects and events, or durable data groups which other researchers, if they were trained, could count, or measure, and come up with the same numbers. In much social and medical research the evidence is not presented in such a way that you can make this decision.

BLACK MAGIC?

Measurement, we have said, can be defined as tying numbers to objects and events according to certain rules. Although we may not be able to explain the rules in detail, most of us can recognize when one of the rules has been broken, particularly if it is a child who is breaking the rules, as in the following dialogue:

You: "How old are you, Kim?"
Kim: "I am five years old and I am seven years old."

While you don't know the name of the rule, you know that Kim has broken some rule about tying numbers to objects. After some discussion, you convince Kim that one age number is enough for a little girl, and she finally agrees that she is five years old and that her brother is seven years old. You then continue your discussion:

You: "Who is the older, Kim, you or your brother? Which age number is bigger?"
Kim: "I am older, my number is bigger."

Once again, even though you may not know the name of the rule, you know that Kim has broken it. Five doesn't come after seven, five goes before seven—anybody knows that—well, anybody who's learned the rule.

On Being Number-Numb

When children break rules in tying number words to objects and events, it is often obvious and usually funny. When researchers do it, it is not so obvious, and can be either funny or disastrous.

It is fascinating to observe how otherwise very competent people often have a blind spot when it comes to using numbers. A faculty member who is exceptionally bright is a case in point. He speaks of numbers as "those ugly little squiggles that are the constituents of a black art." When the abolition of child labor in mine and factory freed children from punishing physical work, he claims, the evil powers rushed in with arithmetic, algebra, trigonometry, and statistics as new forms of child torture. It is his belief that he would still be languishing in Grade I if it were not that a girl named Debbie fell in love with him on the second day of school and, by means fair and foul, smuggled him through the next 12 years of the educational maze. She had a feel for the black arts.

Debbie and my friend parted company after high school, however, and he was left to deal with the mysterious number game alone. The black art followed him wherever he went. Even at parties he would be sought out by one of the high priests of numbers who would casually lean forward and ask, "Are you aware that 1 plus 2 equals 11?" At the mere mention of numbers my friend had learned to generate, instantaneously, a most knowing countenance. He would nod wisely, indicating to those around a profound understanding. The high priest would continue, "Of course, this assumes a base 2 to the number system." My friend would echo, "Of course, of course." If he moved out of the discussion fast enough he was usually not found out. Moreover, if he was able to spot someone who he was certain was also number-numb, usually the hostess, he would repeat the high priest's words, only to have her reply, "Oh, with a base 2 I thought 1 plus 2 equals 2.5." This left him with nothing to talk about other than possible treatments for her stark naked canary, which had lost all its feathers and was crouched on its perch shivering, all covered with shame and goosepimples—his only kindred spirit in all the world.

Why is it that with the mere mention of numbers, for many intelligent people, the lights go out upstairs? We said in Chapter Two that mathematics was a good example of the syntactical level of language. You will recall that this level was concerned with the rules for combining symbols such as numbers. You will also recall that these rules are man-made. Mathematics, then, is like any other game; it consists of players (symbols) and certain man-made rules for combining them. Those of us who like to operate at the pragmatic or semantic levels of language often find mathematics difficult. After a time of experiencing trouble, we get discouraged and say that mathematics is boring. One of the reasons so many people find mathematics difficult is because it was never made clear to them that mathematics is a game composed of man-made rules and man-made players. Unfortunately, some of us tried to play the game using pragmatic and semantic rules. Our plays may have been quite creative, but, unfortunately, did not bring joy to our teacher's heart. We ended up in the corner with five extra pages of mystery to decipher.

If mathematics is not a black art perpetrated on trusting, bright-eyed children, but rather a game of man-made rules with man-made players, what are the rules all about?

NUMBER RULES

At this point we could list and describe some of the syntactical rules for combining number symbols, such as the nominal rule, the ordinal rule, and the interval rule. An alternative procedure is to attempt to discover the syntactical operations by examining a semantic level example of their application. We have noted that measurement can be defined as "rules for tying numbers to objects and events." At the semantic level a measuring instrument is a good example of tying numbers to an object according to certain rules. Therefore, let's examine a man-made measuring tool to discover some of the rules that went into its construction. An ordinary one-foot ruler provides an excellent example.

ONE FOOT RULER

In the case of the ruler, someone has assigned numbers to this flat piece of wood. What rules were followed? Examine the ruler for a moment; notice the symbols; observe how they are placed. If you examine another school ruler, even one made by a different manufacturer, you will find many similarities. While the color of the wood or the paint may differ, the two rulers will have several characteristics in common—the placing of the numerals in both cases has followed the same rules.

Nominal Rule

Notice first that different symbols are used to label different points on the piece of wood. This is the first important rule—the nominal rule: different objects or events are assigned different numbers. If this were not the case we would be faced with a ruler of the type below.

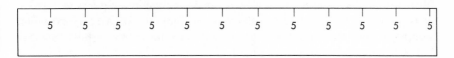

The nominal rule—the naming or labeling rule—demands that you apply different agreed-upon labels (numerals) to different objects and events. In order to be able to do so, people must be able to agree upon and use clear labels, and be able to detect and agree upon differences between objects or events. If the labels are all the same (as with the ruler above), or vague, or if the differences between objects are vague (schizophrenia and depression), it will be difficult or impossible to apply the nominal rule. For example, when numerals are printed clearly, there is very little difficulty in telling them apart. When they are printed quickly, however, "There are 17 schizophrenics in Group One," it is difficult to tell one from another (for example, is the above numeral eleven or seventeen?). Similarly, when numerals are spoken quickly ("seven" and "eleven"), we encounter vagueness, and sometimes confuse one numeral with another. Although vague labels can cause difficulty, it is vagueness in the differences between objects or events (the bundles of data) that cause the most frequent problems. The test of whether we can detect differences between objects or events is to have several individuals independently sort the objects or events being observed into different categories, or, if there are more than one of a kind, into different piles or groups. In this way, we attempt to make the elastic ruler less elastic. This, then, is the basic test of whether we can apply the nominal rule or not. If we can agree on which objects or events go together and which do not, we can usually agree on how to label them.

This first rule—the nominal rule—is considered by some not to be a form of measurement. They point out that it is merely a labeling operation, useful for identification purposes, but that in no way does it tell us anything about "more than" or "less than" relationships, which are supposed to be the heart of measurement. Nevertheless, it can be argued that since counting is a form of measurement, and since it is based on the assumption that we have met the requirements of the nominal rule, it is important to include this rule as one of the rules of measurement. In other words, since the nominal rule is fundamental to all other measuring rules, it should be included. There are many examples, like that facing the psychiatrist attempting to make a diagnosis, in which we fail to meet the conditions of the nominal rule. Nevertheless, we go merrily along counting objects and reporting the results in numbers when, in point of fact, the results are, at worst, meaningless and, at best, highly perishable or personal. This is so when we attempt to count the number of schizophrenics, or early cases of cancer, or anxious people, or good managers, or communists, or improved patients, or members of a social class, when we often have no shareable way of assigning the labels "schizophrenic," "early cancer," and so on.

When different investigators end up with different counts, what are we to assume? Since counting itself is fairly simple, the safest assumption is that the nominal rule has been broken; that is, there is no agreed-upon way of assigning labels to the members of the various populations being counted, and so the counting operation itself becomes a counting of ghosts. Thus, since all other rules of measurement assume that the nominal rule has been met, and, secondly, since there is ample evidence that the nominal rule is often ignored, our personal view is that this rule should be included among the rules of measurement.

Ordinal Rule

Not only are different numbers assigned to different objects and events, but the numbers have a reserved place in the number series— this is the ordinal rule. If it were otherwise, we would encounter such rulers as those below, where different rulers have their numbers in different orders.

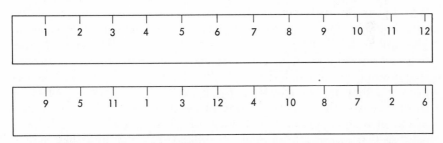

An object measured by the first ruler might be assigned the number 2, but when measured by the second ruler, it would be assigned the number 5. If numbers on a ruler, or in the number series generally, were permitted to play musical chairs, we couldn't use the numbers to talk about order or to indicate where in the series a given object or event occurs. Think of the problems involved in a simple example:

You: "How did Sally make out in the 100 meter free style?"
Paul: "She came in second."
George: "She came in fifth."

In an effort to resolve the conflicting answers, you ask Ringo. He replies, "Sally came in B." Ringo doesn't like numbers and will use them only when no alternative method is available. The three foregoing systems for describing order are outlined as follows:

Paul: 1, 2, 3, 4, 5, 6, 7, 8, etc.
George: 9, 5, 11, 1, 3, 12, 4, etc.
Ringo: A, B, C, D, E, F, G, etc.

It is important to remember that the symbols we use, and where in the series they appear, are man-made decisions. As long as Paul, George, and Ringo use the same rules consistently, they are playing the game. Paul's number series has the advantage of being in common use, large group syntactics; it is blessed by custom. Notice, however, that when George's number series, individual syntactics, is put next to Paul's, there is no contradiction in their replies to our question. In both instances Sally was assigned the position right after the beginning position, that is, position B in Ringo's terms. If George wants to persist in having his own individual ordinal scale, and if he is consistent in its use, we can, if we wish to take the trouble, learn to translate it into our own terms. He is playing by the shareable rules of language, and any one who wants to learn his system is able to do so. If, however, the positions of the numbers change, if George haphazardly changes them from day to day, then this would be a pragmatic system and George would be breaking the "reserved place" ordinal rule.

It should be noted that, in these examples, the nominal rule is violated. You will recall that we stated that, in the case of the nominal rule, different agreed-upon labels are assigned to different objects or events. In our examples, although different labels are applied to differing objects or events, they are not agreed upon. That is, there does not appear to be consensus as to the label for second place—in one case it's 2, in another 5, and in Ringo's system it is B. Each of the rules builds on preceding ones, so that in order to develop an ordinal scale you must first satisfy the assumptions of the nominal rule.

So far we have considered two rules of the number system: rule 1—different objects and events are assigned different symbols or numerals; rule 2—the symbols or numerals are assigned a reserved position in the series of numerals. Notice that the alphabet, as well as the whole number system, fulfills these two rules. Thus, if you want to label events, or simply to talk about their order, the alphabet will do as well as the number system, providing you don't have to talk about more than 26 objects or events. While the alphabet is a useful system for labeling events and describing their order, and while it can be used to describe relations such as "earlier than" and "bigger than," is it really of much use if we want to talk about "how much bigger" one object is than another?

The Interval Rule

Returning to our standard 12-inch ruler, it is apparent that the symbols are placed equal distances apart—a ruler is divided into a series of equal sized units. But how far apart are the letters of the alphabet? One alph? The point is that the order rule makes no assumptions about how far apart the symbols are. All of the following scales meet the ordinal rule:

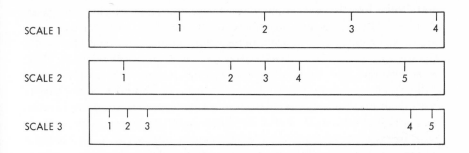

You: "How did our team make out in the 100 meter free style?"
Reply: "We placed first, second, and third."

Does that mean by scale 1, scale 2, or scale 3? Usually when we ask such a question we are not concerned about how big an interval separates the swimmers, but only about the order they came in, regardless of interval. If we want to know the swimmers' times, or the distances separating the best jumps of three pole-vaulters, however, can we handle the problem with only an ordinal scale? Suppose Elvis vaulted 13'9", Blane vaulted 14'0", and Turk vaulted 14'6". How do we communicate this information with only an ordinal scale? We could say Blane beat Elvis by a bit and Turk beat Blane by more than that. So we have communicated

more than just "order" information. We have communicated some distance or interval information as well. How did we do it? By selecting a standard which we called "a bit" (the distance separating Elvis and Blane), and comparing it to the distance separating Turk and Blane, and deciding the latter distance was bigger than "a bit."

The important point, of course, is that to talk about "more than" and "less than" we need to know what a "than" is; if we want to talk about "more than a bit" and "less than a bit" we need to know what a "bit" is.

Selecting a standard, or a basic unit, is a man-made decision. When faced with this problem you look around for a readily available standard and make that your bit. People's feet were usually readily available, and so "one foot" became an early unit for measuring distance. Using people's real feet for measuring distances must have led to certain inequalities as well as inconveniences. When good old Dad's farm was divided, the brother with the biggest "foot" seemed somehow to do better.

Eventually someone with small feet recommended they should have one special foot to avoid arguments; everyone agreed, and of course they decided to use the king's foot. Now a king doesn't go traipsing all over the countryside just to measure things with his foot, so they had to cut one of his feet off, to be sent around for measuring. This left the king with only one foot, hence the origin of the term the "one-foot ruler."[1]

The point is that if we want to talk about how much more than or less than one object is in relation to another, we need a unit, an interval that is easy to apply. It often takes years to develop such a unit and to sell others on using our particular interval, whether it is a second, a bushel, a micro-mercury, a megaton, an ounce, a degree of temperature, a unit of anxiety, intelligence, depression, or a foot.

In summary, if all we want to do is to label or identify objects or packages of data, by using numbers, we follow the nominal rule (different objects get different numbers), and we can do this as long as we can tell the objects, or the qualities of objects, apart. If we also want to describe order relationships between objects, by using numbers, we must include the ordinal rule. We can do this as long as we can order the objects (from earliest to latest, or smallest to biggest), and then we assign the first number (1) in the number series to the first object in the object series, and the second number (2) to the second object in the ordered series, and so on. If, in addition to labeling and ordering, we want to describe with numbers the interval separating objects, then we must use the interval rule and select or develop a measuring instrument which is divided into equal units.

There is one more characteristic about our one-foot ruler that deserves comment. Notice that it has a zero-point. This is so obvious that its importance is often overlooked.

The Ratio Scale

Unless a measuring instrument has a zero point, it is impossible to say anything about how many times bigger or smaller one object, or quantity, is than another. Without assuming that there is such a thing as zero age, we would not be able to say that Harry is twice as old as Mary. Consider the following example: In a test of knowledge of French nouns we have the following results:

Vladimir knew none of the words.
Hamish knew five.
George knew ten.
Gloria knew fifteen.

[1] There may be the odd historian who has some reservations about the complete historical authenticity of this interpretation.

We can portray the results in the following way:

Thus our test of French nouns is a measuring instrument which appears to have all the characteristics of an ordinary ruler—that is, it appears to meet the nominal, ordinal, interval, and absolute zero rules. If this is so, then we can say that George knows twice as many French nouns as Hamish, and that Gloria knows three times as many as Hamish. We can do so only if we are able to agree on where zero belongs on the scale. As you can guess, certain very simple French nouns weren't included on the test the teacher gave—words like *l'amour* and *la bouche*, etc. Therefore, it is quite likely that there are at least five French nouns that even Vladimir knows and that, if included on the test, everyone would get correct. This would involve moving the zero point on our scale five points to the left. Now look at the old scale alongside the new and see how this affects what we can say about how many more nouns Gloria knows than Hamish or Vladimir. In the case of the first scale, we had

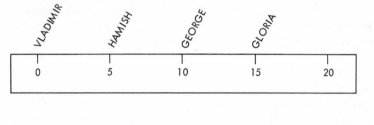

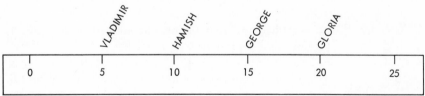

concluded that George knew twice as many French nouns as Hamish, but with the new zero point, this is no longer the case. Similarly, on the original scale, Gloria knew three times as many French nouns as Hamish, whereas on the new scale she knows only twice as many.

It is apparent, then, that if we want to talk about how many times greater or smaller one object is in relation to another, we must have a way of deciding where absolute zero is on the scale. In many cases, such as most test scores—arithmetic, French, intelligence, anxiety, beauty, musical talent—we use an arbitrary zero. In those cases where we are not sure where zero is, we should not: (a) attempt to say how many times bigger or smaller one score is than another, or (b) attempt to use any statistical procedures which involve multiplying or dividing the scores.

You may say that one way around the problem is simply to report that, in the case of our original French test, George knew twice as many of the words as Hamish did on that particular test. In this way you are making no claim about whether George knows twice as many of all French nouns, but only twice as many of the ones on that particular test. And of course you would be perfectly right in doing so. However, most tests are designed to estimate amount of knowledge in a given field, with the test questions representing only a small sample of all possible questions in that area. Thus, while it is relatively easy to say whether someone gets a score of zero on the particular sample of the questions selected, it is extremely difficult to decide whether he would get zero if all possible questions dealing with the topic had been asked.

Most students who succeed in getting poor marks on tests are already aware of what we are talking about. When they say the test was unfair they are saying that it was a bad sample of questions and by poor luck the teacher just happened to select the only questions the student didn't know. Thus, in his eyes, your score of 80 and his score of 10 certainly don't indicate that you know eight times more about the field than he does. In fact, by the time he is finished complaining, he implies that he could have answered any of hundreds of questions about the topic, whereas you were probably able to answer only the eight particular questions the teacher selected. So, in point of fact, he goes away mumbling and concluding that he knows much more about the field than you do, but that the educational people seem almost diabolical in their ability to select those few questions for which he has no answers.

Actually, this is a critical problem facing our educators, particularly with the present knowledge explosion. No longer can we expect professors of physics to know everything about physics, or professors of psychology to know everything about psychology. The professors solve this dilemma by becoming more and more specialized, by carving out smaller and smaller areas or data pools within their own discipline, within which they attempt to become very knowledgeable, and then they dump all that knowledge in the student's lap. How is the student to face the dilemma of information overload at exam time? The naive student attempts to learn all the material that the professor presents, as well as to cover the outside readings. The wise student knows how to study strategically. He has learned to spend more time on some parts of the course material than on others. He has learned to identify the professor's preferences. The student does this on the basis of how much time the professor spends on different topics, by noting what topics appear to excite the professor, by seeing what sorts of questions the professor traditionally asks, by looking over the professor's exam papers for the past years, and by talking to former students. This strategic approach to learning may appear to be unscholarly. Certainly some students carry it to the extreme, devoting almost all their time to attempting to predict the few questions that will be asked, and then applying their very limited energy to studying these few questions. This is, of course, a self-defeating approach to learning. Nevertheless, the serious student facing the impossible task of preparing for all possible questions relies not only on his own particular interests, but also on the biases of the professor in guiding him as to what aspects of the topic require more concentrated work than others.

This discussion is not merely a diversionary bit of advice on the strategy of getting through University. The scientific way to estimate a student's knowledge of a given area would be to have all competent professors in that area write out a list of all conceivable questions. These questions would then be put into huge drums grouped into classes from most important to least important. On a given examination a sample of ten, fifteen, a hundred questions would be drawn out of each drum and would constitute the exam—the number of questions to be drawn depending on the amount of time available for the examination. Such a list of questions would constitute a random sample of hard, average, and easy questions. Under the present circumstances, in which a given professor decides which questions to ask, we have a biased sample of questions covering the field. If a student did repeatedly well on the series of exams based on the random sample procedure, we would conclude that he knows the topic well. When a student does well for a given professor, we are not sure whether he knows the topic well, or knows the professor well. Similarly, when you read the results of a poll concerning who will be the next president of the United States, or on pre-marital sexual relations, are you getting the answers from a cross-section of the population in the country, or are you getting the answers of the friends and colleagues of the person who carried out the poll? In other words, are you learning about the topic, or are you learning more about the biases of the person who conducted the poll? Absolute zeros refer to topics and total populations. Arbitrary zeros refer to samples and biased groups of one kind or another.

In the example of our test of French nouns we were not talking about how many of all (population) French nouns the student knew, but only about how many of those selected for the test (sample). Thus we were using an arbitrary zero, so it is impossible to talk about whether Gloria knows two times or twenty times as many French nouns as Hamish. If we wanted to be able to make such statements with confidence, we would have to test the students on all French nouns. If we wanted to make an approximation, we could test them on several samples of French nouns picked at random from a data pool of all French nouns.

SUMMARY

Nominal scales are used when we compare objects or data clusters and can decide which ones are the same and which ones are different. After this, we can count how many objects fall into each category. Ordinal scales enable us to talk about relations such as more than or less than, or earlier than, or later than. As long as independent observers can rank order objects or events on some less-to-more dimension, they have an ordinal scale.

In some cases, we want to know more than who came first and second. We also want to know by how much one swimmer beat the other. In instances where we want to describe by how much objects or events differ, we use an interval scale. Finally, absolute zero scales are used when we want to compare one event, not merely with another person or sample, but with an absolute zero or population value (knowledge of *all* French words). Each of the four scales, then, has a different purpose, but each succeeding scale assumes that the rules of the preceding scales have been met. If an ordinal scale is to be used, it is assumed that the nominal rule has been met. If an interval scale is to be used, it is assumed that, in addition to the interval rule, the nominal and ordinal rules have also been met.

Since there is still a massive amount to be learned in the social and medical sciences, it is important to remember that nominal and ordinal scales can be of invaluable assistance to us in describing or packaging our data. Just as a crude thermometer is better than no thermometer, so initially crude measures of anxiety, or of management ability, or patient improvement are better than no measure, or casual observation. Furthermore, what may start out to be crude measures are, with experience, gradually refined and transformed into more sensitive measures that can detect smaller and smaller differences and changes.

In the chapter to follow we will have an opportunity to examine some of the challenges involved in constructing simple nominal and ordinal data packaging methods.

chapter 9

Squeezing
Round People
into Square Holes

At one time or another all of us encounter a teacher, professional or otherwise, who provides us with one or two special keys that open doors to rich experience. One such teacher was not a good lecturer in the usual sense. His notes were not well organized into main headings, sub-headings, and clear points under each heading. Perhaps he felt that the world wasn't organized that neatly. When he lectured he was like a man fighting bees. After attacking and being attacked by a topic, he would slump into a chair and say, "Any questions?" When we responded with silence and a studied examination of the table in front of us he would say, "Would anyone ask a question if they understood enough to think of a question to ask?" It was during the next few minutes of his discussion that he would occasionally provide us with one of the special keys—a key question. Such key questions we could carry away with us and use repeatedly and effectively even though the detailed content of his lectures, like the spring snows, gradually melted away.

SPECIAL KEYS

What are the key questions concerning measurement? If you can carve these key questions into your permanent memory they will lead you to the source of much confusion and argument in research as well as in daily life.

Nominal Scales—Do You Have Clear Categories?

Can different observers independently label events and objects in the same way? Consider an example dealing with a nominal scale problem—sorting objects and events in terms of same-different categories. You read an article describing a study of executives in large companies. The researcher wants to know what type of managers are being hired by large firms. He assumes there are two kinds of managers: Group I managers who direct their activities toward maximizing profit and efficiency, and Group II managers who direct their activities to people in the belief that the employees who identify with the company are not only happier, but more productive as well. The researcher sorts the managers he observes into the two categories and reports that out of 90 managers he studied, 60 were labelled Group I type managers and 30 were labelled Group II type managers. Key question: Did the investigator test to see if an independent observer could classify each of the 90 managers the same way? Or, at least, did he see if an independent observer could classify in the same way a sample of the managers—say, 30 of them chosen at random? If not, we don't know if the investigator was playing by pragmatic rules, by potentially semantic rules, or by semantic rules. In other words, how shareable is his data-chunking system?

Almost any time we sort people into a few categories we will find some people don't fit well into any of the categories we have provided. The fewer the categories, the more often this problem will arise. One solution is to increase the number of categories—provide another category for managers who seem to be both Group I and Group II, and another category for managers who seem to be neither (e.g., the boss' son). Or, at least, or in addition, provide a "doubtful" category for those managers who don't fit easily into any of the categories.

By using a larger number of categories you gain two advantages: (1) you don't try to squeeze your data or your people into an inappropriate category; and (2) you usually increase your chances of getting independent agreement among different observers.

Since the conclusions you draw about what type of managers large companies employ depend upon the adequacy and the shareability of the nominal scale you choose—the categories—it is vitally necessary that you provide a sufficient number of realistic and clear categories. This is important whether your nominal scale is concerned with liberals, conservatives, schizophrenics, early cases of cancer, or dominant and submissive rats.

When you test the adequacy of your nominal scale, various results can occur: (1) You may find that independent observers sort few of the objects and events into the same categories that you do. You can then conclude that you don't have a nominal scale at all—crude or otherwise. (2) You may find that independent observers agree in their sorting of some of the objects and events, but disagree about others. You could then conclude that you have the basis of a nominal scale, but that you must clarify the categories by giving more examples or by giving your observers more supervised practice at using the categories. In some cases, you must construct additional categories.

Ordinal Scales

Two key questions are pertinent to ordinal scales. The first and most obvious is—can the objects or events be placed in some order? Suppose we have a pollster whose job it is to determine from a random sample how many families live in rented dwellings, and how many own the home they live in. Is our pollster working with a nominal scale or an ordinal scale? Is a rented dwelling more or less than a house that is owned by the family living in it? We cannot talk in these terms unless we specify number of rooms, floor space, or some other criteria. So we are dealing with a nominal scale and can't talk about ordering these dwellings in any meaningful way. If, however, we talk about number of rooms, we do have an ordinal scale. Three rooms are more than two rooms and two rooms are less than five rooms. Height is another good example of an ordinal scale. We can rank order people, buildings, or other objects in terms of taller than or shorter than.

Having satisfied yourself that you are working with an ordinal scale, the next key question is—how many categories do you need? Too few categories in an ordinal scale can cause trouble, but so can too many. A psychiatrist is interested in evaluating the effectiveness of a new treatment for neurosis. He sets up a five point ordinal scale as follows:

MARKEDLY WORSE	MODERATELY WORSE	NO CHANGE	MODERATELY IMPROVED	MARKEDLY IMPROVED
1	2	3	4	5

IMPROVEMENT SCALE

Category 3 is to be assigned to patients who demonstrate no change; category 4 is to be assigned to those showing moderate improvement; category 5 to those showing marked improvement; category 2 to those who seem to be moderately worse following treatment; and category 1 to those who seem to be markedly worse after treatment.

Our researcher examines the patients before and following treatment, and assigns each to one of the five categories. He has a colleague independently follow the same procedure. You will recall from our discussion of nominal scales that one of the ways of determining whether you have clear categories is to see if different observers can independently label the objects or events the same way. We are essentially following the same procedure here with ordinal scales.

The worst that can happen with the ratings of our two observers is that there is almost no agreement between the two psychiatrists in the assignment of patients to the categories. That is, had the patients been assigned their categories by drawing the numbers out of a hat, rather than having them assigned by a psychiatrist, the results would have been similar. Or perhaps there may be large disagreements between the two. That is, psychiatrist B has assigned some patients to category 5 and psychiatrist A has assigned some of the same patients to category 1, and vice versa. Under either of the above conditions it is apparent that (a) the scale is inadequate, (b) at least one of the psychiatrists has a very personal or pragmatic view of improvement.

What is more likely to result, however, is that the psychiatrists will agree as to which patient should be assigned to category 1 and which should be assigned to category 5, but they will have some disagreements about which patients should be assigned to categories 2, 3, and 4. If there are many such disagreements, you may conclude that you have, in fact, a three point ordinal scale, rather than a five point scale, and in the future will use only three categories.

CLEARLY WORSE	NO OBVIOUS CHANGE	CLEARLY IMPROVED
1	2	3

IMPROVEMENT SCALE

If this is the case, you have an ordinal scale which will detect the results of treatments that have a large effect—e.g., that shift a significant number of patients into category 3. Even a crude scale like this is preferable to no scale at all or, in other words, to individual pragmatics.

But you may decide you want to develop a scale that will be sensitive not only to strong treatments but also to moderate treatments. If so, you devote time to attempting to clarify the distinction between category 3 (no appreciable change) and category 4 (moderate improvement) of your original five point scale. You can attempt to make these categories more distinctive by giving examples of what you consider to be moderate improvement: e.g., "patient may still have severe nightmares, but not as frequently; patient still experiences strong anxiety in presenting class paper, but is attending class more regularly." After attempting to clarify distinctions between the middle categories, you test this scale again and see if you are now getting more agreement between independent judges in the use of the middle categories. If so, you have increased the sensitivity of your scale in detecting moderate change, whereas before your scale was sensitive only to marked changes.

Interval Scales

The obvious question to consider with an interval scale is whether the distance between one point and the next is equal to the space between any two other adjacent points. Suppose we measure the intelligence quotient (I.Q.) of two children using one of the scientifically acceptable tests developed for this purpose. Sonny gets a score of 100 and Jack's score is 130. We test two more children using the same instrument and find Flossie's score is 80 and Bessie's is 110. The crucial question is, although 30 points separate both the boys and the two girls, whether a difference of 30 units at one end of the scale is the same as 30 units at another point on the scale? In other words, does our scale look like this—

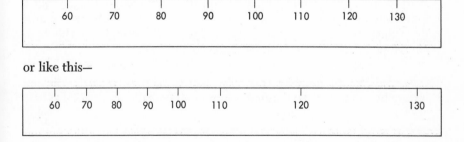

or like this—

In the latter instance, the difference in performance between Flossie and Bessie is much less than the difference between Sonny and Jack, even though the numerical difference is the same.

We face this difficulty frequently in the behavioral sciences when we want to measure such things as moods, attitudes, values, prejudices, interests, aptitudes, opinions, and so on. In such cases, there is no known corresponding physical dimension. We can't measure mood, for example, by using physical tests like liquid displacement (as in measures of volume) or liquid expansion (as in measures of temperature). Thus we are unable to use already developed equal interval scales. If only people who were smarter had appropriately larger head sizes, how much simpler our task would be. Or even if we could discover that they have greater amounts of a particular enzyme or protein in their bodies. So far, no luck—so we are faced with the difficulty of constructing our own scales.

In attempting to construct such scales, we rely on judges to be our measuring tool. In other words, we set people to work sorting our questions or tasks into piles, and instructing them that the difference between one pile and its neighbor should be equal to the distance between any two other adjacent piles. If you can't determine whether the test designer has performed this or a similar maneuver, then you should assume you are dealing with an ordinal rather than an interval data-chunker.

Ratio Scales

When we think we are dealing with a ratio scale, we should ask ourselves the following question to validate this assumption: "Is it conceivable to have a zero amount of this quality?" And if so, is the test using an arbitrary zero, as in the case of the French nouns, or an absolute zero, as in the case of age? These are important issues, since we are not justified in making certain number plays (like multiplying or dividing) unless we have a scale which has a zero point.

One way to tell if a particular measuring instrument has an absolute zero is to see if all members of the population are included or if only a sample was drawn. In the case of most examinations, measures of attitude, value, intelligence, popularity, and so on, a sample of the total potential population of descriptive statements is used.

It should not be assumed that scales with arbitrary zero points are to be avoided or are useless. The only danger with such scales lies in their use by unsophisticated, number-numb individuals.

REFINEMENT

The development of increasingly sensitive measuring instruments in the behavioral and medical sciences is a critical need. The use of pragmatic scales or of unnecessarily crude nominal or ordinal scales, except in the very early stages of investigation, is lamentable. Such practice wastes the time and effort not only of the researcher involved, but also of other researchers who follow his lead. While the use of pragmatic scales often falsely gets our hopes up and leads us down blind alleys, the use of unnecessarily crude nominal and ordinal scales can result in rejecting a treatment which is potentially useful. For example, the psychiatrist who is using a crude three point scale to test the value of a new treatment for neurosis that leads to marked improvement in only a few cases, but to moderate improvement in many cases, will not be able to detect the moderate changes and, therefore, a potentially useful treatment is rejected. There is a great need for creative researchers who can devote their ingenuity and energy to the development of shareable nominal and ordinal scales of ever-increasing sensitivity.

Up until now, we have stressed the point that measuring scales should be shareable. In addition, they should provide us with useful information. The inclination to tie numbers or labels to mundane or misleading information is great, as we shall see in the next chapter.

chapter **10**

Who in the World Are You Talking About?

Now that you have some rough idea about how to measure, whom or what are you going to measure? That obviously depends upon who or what you want to end up talking about.

Suppose you want to end up measuring and talking about someone you know very well. Think of someone specifically. Immediately you can apply a few nominal scales.

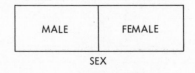

MALE	FEMALE

SEX

An independent observer would probably agree with whichever classification you check, so that we have met the first criteria of applying the scale: independent observers agree.

Now consider hair color:

BLOND	BROWN	BLACK	RED

HAIR COLOR

You consult with an independent observer who suggests that you need an additional category between blond and brown. So you change your scale accordingly.

BLOND	DIRTY BLOND	BROWN	BLACK	RED

HAIR COLOR

Now let us apply some ordinal scales:

DULL	AVERAGE	BRIGHT

BRAINS

The independent observer suggests that you include an "above average" category. In order to accommodate your friend, again you agree to modify the scale accordingly:

DULL	AVERAGE	ABOVE AVERAGE	VERY BRIGHT

BRAINS

Now apply another ordinal scale to describe your friend's height:

SHORT	MEDIUM	TALL

HEIGHT

If you have a tape measure you can use an absolute zero scale (ratio scale) to describe height.

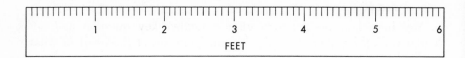

Completing these scales was probably not too difficult for you. What about the following?

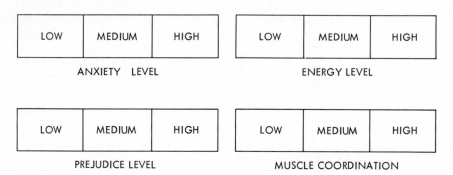

These scales are probably not as easy to complete. Why? While people's height and sex do not change (too markedly at least), an individual's anxiety level, or energy, or prejudice level changes from month to month, day to day, hour to hour, or even minute to minute. To be able to complete the scales, you are forced to leave out a lot of information about your friend. You may want to write in such comments as:

"It all depends. Sometimes he is anxious, sometimes he's not. Sometimes he's lazy. If you force me to complete these ratings you're going to end up with a biased picture because you forced me to leave out so much information. You've assumed that anxiety and energy and prejudice are like height, unchanging, and of course that isn't so. If you want me to complete these ratings realistically, you will have to let me qualify my answers to apply to certain times like this morning, or most Monday mornings, or most Friday nights, or after he loses a bet. Otherwise I'm giving you a biased picture—a fictional average that tells practically nothing. You are kidding yourself if you think you're learning much about my friend. If you describe my friend on the basis of my overall ratings to someone else who knows him well, their reply would be 'Who in the world are you talking about?' Most of the nominal and ordinal scales you have given me may apply to the odd person who is anxious almost all of the time, or energetic almost all of the time, but in my experience, there aren't too many people like that, so I don't think your scales are very useful. Therefore I won't fill them out unless you let me qualify my ratings by stating the particular times, or occasions, I am thinking about when I describe my friend."

Now if you were to protest in this way you would be justified in doing so. You have laid the groundwork for another key question and, although you may not know it, the key question lies at the heart of what is called sampling theory:

WHO, OR WHAT IN THE WORLD ARE YOU TALKING ABOUT, AND AT WHAT POINT IN TIME, OR UNDER WHAT CIRCUMSTANCES DO YOUR DESCRIPTIONS OR RATINGS APPLY?

This key question should be asked whether you are talking about your friend, or a test of French nouns (knows what French nouns when?), or voting preferences, or pre-marital relations. The key question condenses to: what, or who, and when? One way of handling this key question is to specify within your categories the circumstances and the time at which ratings are to be made.

Obviously, we can't keep measuring everything or everyone all the time. But we can and should limit our conclusions to our data samples, to the people or things measured, and to the times at which they were measured, unless, of course, we have hard evidence that what we conclude applies to the same person at other times. Otherwise we are misleading the people to whom we report our results.

When a counseling psychologist reports to the press that pre-marital relations are rampant on the campus, who is he talking about, under what conditions, and at what time? Is he talking about the whole student body? Or is he basing his conclusions on interviews with five or six students who are in trouble and using self-justifying hearsay evidence about the other eight thousand students. In other words, is he filling out a nominal scale for all students on the basis of hearsay evidence from questionable witnesses, and with no attempt to specify frequency or to distinguish between the highly promiscuous students, or those engaged to be married at the end of term, or those who are not participating in such relationships?

But, you may say, such information is hard to obtain. All the more reason to suspect the conclusion of experts about hidden, distant, or complex behavior. When it is difficult to obtain information repeatedly from the group you want to talk about, you should approach the conclusion cautiously. A good investigator will report the number of people sampled, and will specify under what circumstances and for what time his conclusions apply—just as you wanted to do before completing the ratings on your friend.

We said previously that the key question: "Who, or what, was measured, when and under what circumstances?" lies at the heart of sampling theory and of much that goes under the heading of statistics. While you don't have to know all about sampling theory or statistics to ask the key question, some knowledge will be helpful.

FIGURES DON'T LIE, BUT . . .

There is a saying that figures don't lie, but that liars do figure. It is interesting that statistics has acquired an unsavory connotation as if it were, by design, a tool for deception.

"Nine out of ten doctors prescribe Calmies."
"You get 24% more juice with the Great-Embrace orange squeezer."
"Aunt Sadie won 100 games of Solitaire in a row."

Did you find yourself saying, "Yes, but I'll bet Aunt Sadie cheats?" Certainly you can win, if you break the rules. Of course, if you don't know the rules, you can't tell if Aunt Sadie is a cheat or a genius. Like Solitaire, statistics is a game with rules. Break the rules if you want, but the game is no longer statistics, any more than the game Aunt Sadie plays is Solitaire.

What does it mean when the velvet-voiced announcer says: "Nine out of ten doctors prescribe Calmies?" Does it mean that 90% of all doctors in the world made a sworn statement before a lawyer that they prescribed Calmies regularly? Think of the insurmountable task involved in getting this information. Or does it mean that the advertising firm kept phoning groups of ten doctors until they finally located one group of ten in which nine said they prescribed Calmies? Or does it mean that a script writer, over his martini lunch, simply created the statistic knowing full well it would take too much work for anybody to disprove the statement?

Perhaps we should not be blaming statisticians, or Aunt Sadie, or the announcer, or the script writer, but rather ourselves for not knowing enough about the game to be able to spot the work of a cheat or an amateur or a fool.

Statistics can be a valuable tool of communication when it is used as a shorthand method of talking about the semantic world. There are two main branches of statistics: (1) descriptive statistics, and (2) sampling statistics.

When you say "There are 15 girls and 17 boys in our class," you are using descriptive statistics. There is no guesswork involved in this statement, provided that: (a) you can tell a girl from a boy, (b) you can count, and (c) the whole class is present when you make your count. *Descriptive statistics*, then, is a shorthand method of making a statement about some observations when you are able to observe the whole group you want to talk about. It is a simple way of chunking data into neat, easily managed bundles. But many times it is not possible to observe or to count everyone in the group you want to talk about. If it is a very large and widely scattered group, it would cost too much in time and effort to observe or count every member. This would be the case if you wanted to make a statement about how many doctors prescribe Calmies. So you can either forget all about making any statement, or you can try to make an estimate of how many doctors prescribe Calmies. *Sampling statistics* and sampling theory provide guides for you to reduce the error in your estimate, and help you to use your time and effort most effectively when you can only measure or observe a part of the group you want to talk about.

You already know something about sampling theory. You know that, generally speaking, the more doctors you sample, the more accurate your estimate will be. You also know that the wider the cross section of doctors in your sample, the better your estimate will be. For example, if you question only public health doctors, even if you contacted all public health doctors in the world, you would be getting a biased sample, since very few public health doctors prescribe any drugs at all. Therefore, if your purpose was to discredit Calmies you could question only public health doctors and report that "only one in 1000 doctors prescribe Calmies." If you are honest, you would say that "only one in 1000 public health doctors prescribe Calmies." If, on the other hand, you wanted to promote Calmies, you might include in your sample only those doctors who hold stock in the Calmies Drug Company, and who, therefore, possibly prescribe Calmies for many ailments. Then in your advertisement you may be able to say that "99 out of 100 doctors prescribe Calmies," when you should say "99 out of 100 doctors who hold Calmies stock prescribe Calmies."

It is on questions like these, involving sampling statistics, that the experts and the public must rely in making many of their crucial decisions. Shortly we will consider a list of some of the important questions facing us, the solutions to which rely on the adequacy of our sampling statistics —who or what was measured when?

Before doing this, let us consider again the problem of determining how many males and females there are in a given class. By so doing we will get a feel for the kinds of questions we should keep in mind when we are applying sampling statistics. Let us assume, just for the sake of argument, that we don't want to bother counting all of the people in the class, so we decide to take a sample and use this as a basis to estimate the number of males and females in the class. There are various ways we can go about it.

If we do not want to wait until the whole class is assembled, we may decide to use as a basis for our estimate the number of males and females who are in the classroom five minutes before the class starts. Let us suppose that we know there are 40 students in the class and that when we go in five minutes before class starts we find there are three girls and one boy. On the basis of this sample we may conclude that there are three times as many girls as boys in the class, or 30 girls and 10 boys. This may turn out to be a fairly accurate estimate, or it may be that girls tend to arrive before class begins more often than do boys. If this is the case, then we have a strongly biased sample, not a good cross-section, and we will end up with a large error in our statement about the number of females relative to the number of males in the class.

Or we may take another approach to getting our estimate. We may wait until the class has assembled and the lecture is under way. Since we do not want to interrupt the class, we merely look through the window in the door. In this way, without pressing our noses to the glass and distracting the students, we can see the first two rows clearly. Let us assume that there are five rows of desks with eight students per row. As we look through the window, we find that there are 16 boys in the first two rows, eight in each row. Thus we assume that since our sample consists entirely of boys, this must be an all male class. It may be, however, that the boys prefer to occupy seats near the door so that they can get out immediately after the lecture is over, and that had we pressed our noses to the glass, so as to get a count of the whole group, we would have found that there was an equal number of girls and boys in the class but that the girls were concentrated in seats away from the door.

Thus it is important first of all to realize that there will be some error in every estimate based on a sample. Second, the size of the error will depend on the size of the sample. Third, the size and type of the error will depend on whom we include in and whom we leave out of the sample.

Now let us consider examples of the kinds of statements that we want to make, and in which the accuracy of the statement is dependent on how well we apply our sampling methods:

1. (a) Percentage of people in the United States who plan to vote Democrat in the next federal election.
   (b) Percentage of people who plan to vote Republican.
   (c) Percentage of people still undecided.
2. (a) Percentage of cavities found in people who use Crest toothpaste.
   (b) Percentage of cavities found in people who use Control toothpaste.
3. Percentage of people who watch N.F.L. football on T.V.
4. Percentage of people who plan to buy a new car this year.
5. Percentage of patients who improve on drug X.

We have said that when we cannot observe or measure each member of the group or population we want to talk about, we can observe or measure a sample of them and at least get some kind of answer to our question based on the sample estimate. Some of these estimates based on samples are accurate. "Kennedy will beat Nixon." Other estimates are inaccurate: "Dewey will beat Truman."

In these two cases we have a chance to test the accuracy of the estimates—based on samples—since on election day almost the entire group turns out to declare its hand.

What are some of the reasons for inaccuracy in our estimates based on samples? In the 1948 American presidential election the major polls (Roper and Gallup) picked Dewey over Truman. The major reasons for error in polls are the same ones which operate when you attempt to find out whether your crowd is going to the weekend football game. The reasons for error include: (1) you don't usually poll every member, and (2) some members change their minds, or are undecided at the time you question them. You can increase the accuracy of your estimate of how many of your friends will be at the game by questioning every member of the group, and by questioning them several times right up to and including the afternoon of the game. But, you may reply, and rightly so, that such a ritual would require much time and effort. Usually you don't need to know precisely how many of your crowd will be at the game. Probably you just want to be sure of some company, or of a few specific individuals. If you are responsible for getting the tickets, however, you want a more precise estimate because the cost of error will be high. Similarly, in a public opinion poll, if the cost of being wrong is high, and if it looks as if it is going to be a close race, you have to be more precise in your estimate. This means larger and more carefully selected samples of people who are questioned several times right up to the time of the election.

Prior to 1948, the polls were usually accurate not because particularly precise procedures were used, but rather because the elections were one-sided, and even crude methods were able to pick out the voting preferences. In 1948, however, these crude methods were no longer adequate in such a close race.

It is apparent, then, that every estimate based on a sample will be in error by some amount. Some methods of sampling lead to larger errors than others. In fact, it is possible to estimate the size of the error that specific sampling procedures will likely produce. For example, by questioning a given size sample three times prior to an election you may be very certain that you can estimate the percentage of the population voting Republican to within five percent. By increasing the size of the sample, and increasing the number of times which the members of the sample are questioned right up to the day of the election, you may be able to reduce the error to two or three percent, but at high cost. If the election is not close, if the difference between the two parties is greater than five percent, then the pollster will likely be accurate with the former procedure. If it is a small difference, however, say less than two percent, then the pollster may back the wrong horse.[1]

IT'S A CHANCY BUSINESS

The kernel of the sampling problem is simply this: because we can never get all the information about any question, we must, of necessity, rely on the bits of information we can get. Since we can never get the whole story, we must rely on parts of the story to reach a conclusion.

This problem faces all of us—doctor, lawyer, Indian chief, mother, father, son, daughter, teacher, employer, employee, congressman, president, spy, pollster, lover. Because we operate most of the time on partial information dipped from the data stream surrounding us, we realize that there is a good chance we will be wrong to some degree. Our everyday speech indicates that we are aware that decision making is chancy. One teenager says to another, "What are the chances of getting your father's car for the dance?" One congressman says to another, "What are the chances of getting this bill through Congress?" The prospective sponsor of a new T.V. series wonders, "What are the chances of the new series drawing more viewers than N.F.L. football?" The employee speculates, "What are my chances of getting a raise?" A company executive wonders, "What are my chances of avoiding a strike if I don't go along with the union's request?" A president and his cabinet ponder the chances of a naval blockade of Cuba leading to war.

The teenager, on the basis of the bits of information that he has, may decide that the chances of getting the family car are probably good. The bits of information he has include: (1) his parents usually stay home Saturday night, (2) he doesn't think he has done anything that week to antagonize his father, and (3) his father seems to be in a good mood. The son attempts to increase his sample of information by subtly questioning his mother—"I hear there's a good T.V. program on Saturday night. Are you and Dad going to watch it?" Questioning can be too subtle sometimes. The mother replies that they do plan to watch T.V. Saturday

---

[1] For those who want a closer look at polling and sampling procedures, see Bernard Berelson and M. Jaanowitz, eds., *Reader in Public Opinion and Communication* (Glencoe, Ill.: The Free Press, 1953), pp. 584–93. Also consult C. Schettler, *Public Opinion in American Society* (New York: Harper and Row, Publishers, 1960), pp. 465–92. Also see J. Coleman, E. Heaw, R. Peabody, and L. Rigsby, "Computers in Election Analysis: The New York Times Project," *Public Opinion Quarterly*, XXVIII (1964), 418–46. In the same issue of that journal consult L. A. Marascuilo and Harriott Amster, "Survey of 1961–1962 Congressional Polls," 497–506.

night, but she neglects to mention that the reason they are going to watch T.V. at home is that the car will be in the garage for repairs on the weekend. The son is now convinced that getting the car is a sure bet and tells his friends the transportation problem is solved. He made the best decision he could with the sample of information he had. Why hadn't Harvey used the direct approach to his father, or at least simply said to his mother, "What are my chances of getting the car Saturday night?" This seems like such an obvious course of action; why wasn't it followed? Because, like the rest of us, this teenager has learned that information is not always like a can of beans, immune to the way it is picked up. Some information is more like a snowflake; grasp for it and you change it, or lose it. Let it float gently onto your glove, and you have a snowflake in all its beauty and intricacy. From experience Harvey has learned that the timing and phrasing of a question can affect not only the answer he gets, but also the respondent's subsequent behavior, making him in some instances defensive or vigilant, and in other instances open and relaxed.

Pollsters and social scientists, too, have problems similar to those facing Harvey—when and how you ask your question can affect the answers you get. In some cases information can survive rough handling: e.g., name, address, telephone number, rank, and serial number, but in many cases the information is more like a snowflake, or at least like plasticene, and is very sensitive to the handling it gets: e.g., information relating to certain aspects of religion, politics, personal hopes and fears, or sex.

Because so many crucial personal and institutional decisions depend on samples or bits of information, it is increasingly important to appreciate how statistics are used to help us move from bits of information to broad generalizations. In the next chapter we will examine this aspect of statistics.

chapter 11

Leaping to Conclusions

The main purpose of statistics is to help us leap to conclusions with the help of numbers. Leaping to conclusions is necessary when we must make a decision on the basis of limited information; we may toss a coin, say "eenie, meenie, minie, moe," consult a friend, or look for trends in our past experience to guide us. Statistics is nothing more than a ritual to help us locate trends or leads in the partial information we possess which will guide our decision making.

Suppose you plan to consult a friend about a decision you must make for which you have only limited information. Whether your friend will be able to come up with a good bet for you depends on (a) the amount of information you can give him, and (b) how durable it is. The same holds true with statistics. There is no magic about it. The more durable and representative the information you feed into a statistical ritual, the better your chances of getting a good bet out of the ritual. Feed in too little data, perishable data, or biased data, however, and you will get back neatly-wrapped, sweet-smelling garbage.

The value of statistical procedures is that they can often assist you in locating trends or patterns in your data that you might otherwise overlook. Statistical procedures, in a way, help to magnify or to make more noticeable patterns in the information you have. So, by following the rules of the statistical game, you can often determine whether there are trends hidden in your limited information. But break the rules of the statistical game, and you can fool yourself, and perhaps others, by reporting phoney results. When this kind of hanky-panky happens, don't blame statistics; blame the person who broke the rules of the statistical game. You don't blame the game of Solitaire because Aunt Sadie "wins" every game, you blame Sadie, or perhaps say that it's not standard Solitaire but "Sadie Solitaire." Similarly, through ignorance or a desire to win, there are people who play "Sadie Statistics" as well. That's why if you are going to rely on statistics—your own or those of others—it is wise to learn some of the basic rules of the statistics game before relying heavily on conclusions or recommendations manufactured with the help of statistical tools.

GENERALIZING FROM SAMPLES

In research, as in life, data comes in bits or samples—samples of ore, public opinion, radioactivity, blood, or cigarette smokers—and researchers are not so much interested in the samples, as in what the samples can tell them about the bigger parts from which they were taken. For example, in geology a researcher may have a theory about where a particular copper-bearing strata lies, how deep it is, and where its boundaries should be. Between the geologist and his theoretical copper mine, however, lies a barrier of earth. Rather than go ahead and move tons of the earth curtain, the researcher will probably drill sample holes at various places and to various depths. He is, in a sense, taking shots at his hypothetical mine and determining whether he makes any hits.

Similarly, a surgeon may suspect, or have a theory, that you have lung cancer, but, rather than move away a great slab of your flesh, he will probably probe for samples of tissue and see if the sample bits show signs of cancer, just as the geologist will see if the earth bits show signs of copper. The samples may reveal nothing. This does not mean that there is no copper deposit there, or no lung cancer. In both cases the samples may have been poor representatives of what surrounded them; that is, if the investigators had gone just a bit to the left or right, or a bit deeper, or taken more samples, they may have come up with information that supported their hypothesis. Just because your sample shows nothing doesn't mean there is nothing there, but how many samples can you afford or tolerate?

Look Again

Research is like a love affair. The ingredients include: (1) your image of the girl; (2) the real girl as she would appear if you were all-wise and had access to all information about her; and (3) the bits, pieces, or samples of information you have, some of it clear, some of it vague, some of it twisted by memory or biased senses. You have a picture, of sorts, of what the girl is like; gradually you collect more bits or pieces of experience and fit them together. Some pieces are hard and clear and fit well into your picture. Some are vague and soft and can be made to fit your image. Others are hard, and at first sight do not fit into your picture at all. With time some of them can be molded to fit or filed away out of sight. But if the pieces are big enough, hard enough, and cannot be filed away, but keep turning up repeatedly, they can lead to little or big changes in your picture. Changing a once-loved picture is a very painful process, and we know the degrees to which a lover will go to ignore, twist, and blink away negative data. The researcher, too, once hooked on a project, develops a picture of the object of study. He gradually accumulates bits or samples of information, testing them to see how well they fit his picture.

Just as you learn only a little about a girl on one date, so you learn only a little about an object of study in one experiment. If the date was a quick coke after class, you learn little of what the girl will be like at a formal, on the ski slope, in bed, as a mother, as a cook. In fact, you learn very little about what she's like for a quick coke because you don't know what she would be like for a second quick coke date, when it's not a first date, when she's tired, when she's menstruating, when she's just failed an exam, when she's seen you out with another girl just before, when the star quarterback is at the next table, when you're tired, when you're broke and can't take her to the prom, when the campus beauty queen is at the next table, when you're excited about winning a scholarship, when you're completely happy just to be there with her and therefore not repeatedly casing the joint to see who else just came in.

The foregoing is what sampling theory is all about, about what you already know—that you don't get to know a girl on one date, or a person on one occasion.

The more samples of information you take and the bigger the samples of information you take, the more likely you are to get a durable picture of the girl, and the better will be your chances of predicting her behavior. In research, too, the more samples you take under a variety of conditions, the more likely you will be able to predict the rat's behavior, the astronaut's behavior, the gas's behavior, the dollar value's behavior, and so on.

If you keep in mind the question of what you learn about a girl or a guy on one date, you will have no difficulty in understanding the essence of research. It is an effort to construct a durable pattern of information from bits, or samples of data. There is no magic, and there is no guarantee that your predictions from one sample will be correct for the next sample you collect. Although good researchers take more care than poor ones in their data collection, recording, and analysis, and in planning their next experiment so that certain kinds of information can be squeezed or distilled out of the data, there is no magic that can replace the value of large samples and repeated samples, but how many samples can you afford or are you prepared to bother with?

Replication—A Red Herring? We have stated that the scientist's search is for durable information. One device to test for durability is to repeat your own or another investigator's experiment—which is like taking your latest love out for a second quick coke date. This is the principle of replication, and while an extremely important principle, it is far from simple and becomes increasingly fascinating the more you play with it.

The term replication implies an exact repeat in detail of all the experimental conditions. But you will say, "Ah yes, within reason—that doesn't mean the experimenter has to wear the same shirt he did for the earlier experiment," and you would probably be right. But does it require he run the experiment at the same time of day, the same time of year, with the same janitor cleaning out the animal cages, with the same lab technician, with the same instruments (now a little older, of course), and so on? You see, when another investigator cannot get results similar to yours, the first thing you look for are differences in his experimental procedure, and you can always find them—some perhaps farfetched. However, the differences which are reasonable to ignore today often become an important experimental variable tomorrow. A few years ago no one was particularly concerned about whether white rats were petted between experiments. Now there is evidence that such petting influences them in a variety of ways that affect the experimental results. A few years ago no one was concerned about who did or did not wear a watch with radium painted on the dials, or whether the girls painting the dials licked their brushes. Such events could be ignored, and yet such innocuous licks were silently and slowly translated into lethal doses of cancer.

The point to be made is that (a) since it is impossible to repeat or replicate an experiment in every detail, and (b) since we have no sure way of knowing when a small difference in detail is important, we should do away with the concept of replication and replace it with one that is less misleading. We want concepts that indicate that we are interested in the shareability of the experimental results being reported—concepts like exportability and perishability seem appropriate.

Ordinarily, it is almost impossible to call an investigator a liar, even though you attempt to repeat his procedure in as much detail as you can afford time and time again, and still fail to get results similar to his. Nevertheless, you can rate most experimental results on their exportability. Some experiments can be exported to any normal adults, some experiments can be exported to the average specialist in a given field, some can be exported only to a select group of specialists trained by the master, some experiments can only be exported to believers in a given theory.

Rather than asking the question, "Can this study and the findings be replicated?" let us ask to what extent it is exportable—what restrictions are placed on its exportability—how perishable is this research finding? Or how durable is the finding? If the exportability of a given experiment is low, you can speculate as to why—is it the experimenter's *unique touch* on the patient, on the apparatus, on the rat, on the raw data, on the statistical procedures, on the calculator or computer, on the English language, on the journal editor?

Sample Rules

When someone reports that statistics prove that Crest toothpaste reduces cavities, you can now translate such a statement. Someone has used some statistical rituals to look for trends in his samples of data. If the information he collected is durable, and if he didn't break any of the rules of the statistical game, then Crest looks like a good bet over Brand X. But you can't put much confidence in the findings if he has broken some of the rules of the statistical game. For example, he may have excluded some of the data from the analysis or report: e.g., the results of some Crest children who ended up with a lot of cavities. Or he may have set up two unevenly matched groups of children by putting more children who were regular brushers into the Crest group than in the control group. Thus this group could end up with fewer cavities because of more brushing, rather than because of Crest toothpaste.

Statistical procedures are usually based on the assumption that the two groups were equal to start with. This assumption is usually met if we set up the two groups to be studied by using a randomization procedure. For example, the names of all the children to be studied are put in a hat. The first name drawn goes into the Crest group, the second into the Brand X group, the third into the Crest group, and so on. In this way there is no reason to assume that more children who are regular brushers end up in one group or the other. It may happen by chance, but that is a risk researchers must take. If they repeat the study and get similar positive results for Crest, however, we have increased confidence since it is unlikely that in two successive experiments using the randomization procedure we would end up with more children who brush regularly in the Crest group. This is one of the reasons why being able to repeat your own experiment, or someone else's and obtain similar results is so reassuring.

Failure to use randomization procedures is one of the common infractions of the rules of the statistical game; a second common infraction involves not including all data in the analysis. For example, suppose we want to do a soap commercial in which women judge which brand of detergent does the whitest wash. Let us say that there is not a difference between the two brands, so that we find that 25 percent of the time women choose the pile washed in Brand X, 25 percent of the time they choose the pile washed in Brand Y, and 50 percent of the time they see no difference. We film the tests and show on TV only those in which our brand is selected. By leaving information out, we are able to report in favor of our own brand. This is obviously like cheating at Solitaire. There are instances, however, where the researcher, perhaps in good faith, leaves data out of the analysis and, by so doing, is able to demonstrate a significant effect which, had he included all the data, would not have been the case. For example, the researcher, after looking at the data, may notice that three children in the Crest group show a very large number of cavities. He contacts the mothers of these children and inquires about their eating and brushing habits. Perhaps he finds that all three of the children drink a great deal of pop and eat pounds of candy. He then decides that these are atypical children and takes their data out of the analysis. Now, the Crest and the control toothpaste groups differ by more than the margin set up to provide for chance differences. But notice—the experimenter has not checked with all the children in the control group to find out whether there are an equal number in that group who drink a lot of pop and eat candy. The ground rule that should be followed in any instance in which data is being left out is that the data should be reported in the study as an appendix, and that the investigator has an obligation to justify why he took these data out.

In this way other investigators reading the report can decide for themselves whether the omission of the data was warranted.

SIGNIFICANT DIFFERENCES

We have said that the purpose of statistics is to help us identify trends or differences in our data that are worth advertising to the public or to other researchers. It is important to note that in every study using a before-and-after method (O_1 **X** O_2) there will always be some difference between O_1 and O_2. Similarly, in using the control group method,

$$O_1 \ \text{X} \ O_2$$
$$O_{1a} \qquad O_{2a}$$

there will always be some difference between O_2 and O_{2a}. The problem, then, is not just obtaining a difference, it is to come up with a shareable method to help us decide if the difference is worth shouting about, or publishing. We need a decision aid. For example, if in the toothpaste study we found that the Brand X children ended up with a total of 400 cavities, and the Crest children with 395 cavities, we have a difference, but hardly one worth shouting about. The real question is how big a difference do we need before we conclude that it is a significant difference?

Statistical procedures provide us with a shareable method of estimating the range of differences we expect between two groups even if there is no difference between Brand X and Crest. For the sake of discussion let us assume that we have carried out the appropriate statistical calculations and have found that differences of up to 50 or 60 cavities can be expected even if the two toothpastes are identical. Thus differences of this size between the two groups can be expected as a result of the selective operations of one or more of the four rogue suspects (in-the-gap, time-tied, elastic ruler, on-stage) on one of the groups more than on the other. We can never be completely sure that the four rogues are operating to exactly the same extent on both groups, we can only design our experiment in such a way that it is likely that the four rogues will affect both groups approximately the same way. Then, with statistical procedures, we allow for a margin of difference that can be blamed on the rogues, and only when we exceed this margin of difference do we consider the possibility that our treatment X is having an effect over and above the effects of the rogues.

It is important, then, to appreciate that we can obtain a significant difference, beyond our margin difference, under a variety of circumstances: (a) if our treatment X is having an effect over and above the effects of the rogues; (b) if one or more of the four rogues is having an unusually strong effect on one group over the other; (c) if the groups were not equal to start with; (d) if the researcher or advertising agency has left out some of the data; (e) if there is a mistake in our calculations or a fault in the computer. When we obtain a significant difference, how do we decide to have confidence that it was the treatment X that led to the difference? To the extent that we can answer the following questions positively, we gain increasing confidence that the treatment X was, in fact, a key factor, a durable first-class suspect.

1. Does the investigator demonstrate that he has made a careful attempt to control the four rogue suspects? Yes_____ No_____ For example, you would have more confidence if a control group design had been used than if a before-and-after design had been used.
2. If two or more groups were used, was there a reasonable attempt made to assure that they were equal to begin with (i.e., randomization)? Yes_____ No_____
3. Did the investigator use enough people in each group to make you feel that his samples adequately represented the target population he wanted to end up talking about (e.g., kids of different ages and from different socio-economic backgrounds)? Yes_____ No_____
4. Did the investigator publish or make available his raw data so you could check his calculations or data packaging procedures? Yes_____ No_____
5. Did the investigator repeat his study and get similar results? Yes_____ No_____
6. Is he an established investigator whose work has usually proved to be durable in the past? Yes_____ No_____
7. Has another investigator repeated this study and published similar findings? Yes_____ No_____
8. If so, was the second investigator independent of the first investigator (i.e., not his graduate student or employee)? Yes_____ No_____
9. Do the findings make sense in terms of other durable findings in the same field? Yes_____ No_____[1]

In summary, all statistical procedures do is identify any trend in the data that the investigator chooses to include in his analysis. This trend may be there because of treatment X, or because of the way the investigator, wittingly or otherwise, built it into the data. It is only through an examination of the research methods he employs that you can decide how much confidence you have that the data pattern or trend is related to treatment X.

[1] Too much reliance on item 9 above will serve to reduce the possibility of new and startling information being introduced into the literature. Therefore, if most of the other questions are answered positively, item 9 should not be used as a basis for denying the investigator his right to publish.

four

The
Scientific
Establishment

SCIENCE is big business. The day of the struggling young researcher operating on a shoestring budget and making do with chewing gum and rubber bands is over. Not only has science gained in status as an enterprise worth supporting, like many large institutions it has developed rituals, rules, and procedures which have become ingrained. The professional scientist has had to learn and adjust to these.

Like the housewife, the successful scientist must, of necessity, become a multiskilled individual if he is to survive. He must be alert to fiscal policy, and market changes, and even foreign affairs, since these factors will affect his ability to compete successfully for a slice of the available research funds. He must compete with other organizations in recruiting a good staff. He must also compete for space and equipment. He must be an able administrator so that he can maintain his staff and meet their professional needs. Most important, he must do research and publish his work. If he fails to publish, then funding, staffing, and equipping all become more difficult, to say nothing of his own professional advancement. In the atomic age, the scientist with the peptic ulcer is perhaps the norm. The big business aspect of science is covered in Chapter 13.

The main content for Chapter 12 is the thesis or doctoral dissertation. This is the one almost universally-employed procedure for professional training in science. This chapter is not meant to describe how a thesis should be written, nor is it meant to ridicule what can be a learning experience of great consequence. Rather, we hope to portray the thesis preparation not only as an heroic undertaking, but also as a human maze. Thus, in addition to brains and drive, the student should have the ability to pause occasionally en route to appreciate that he is in the midst of scientific puberty rites. From many rites he will no doubt benefit, while others may seem quaint, obsolete rituals, designed to provide employment for the medicine men.

chapter 12

Training

In this chapter we will focus our attention on the beginners in the science game and look in on their practice sessions. The practice court for the budding scientific player is the university, college, or high school, and his progress through educational hurdles, or loopholes, constitutes the drills and calisthenics of the science game.

The importance of education as a prerequisite for a scientist can hardly be overestimated. The association between high levels of education and scientific work is a close one in the minds of most people, and this belief will likely endure. It is increasingly rare for the untrained or self-educated man to find a niche for himself in scientific endeavors.

What are the essential requirements for being accepted by the public at large as a legitimate scientific player? A private source of funds? An interest in science? A bright or enquiring mind? A score of 140 on an I.Q. test? A father who is a university president? None of these—the essential and sufficient condition for being allowed off the bench and on to the scientific playing field is a series of alphabetic characters, usually Ph.D., which are found after an individual's name. Unfortunately, those interested in playing the game cannot legally assign these characters to themselves, and it is becoming increasingly difficult to obtain them by sending in box tops and a dime to good old Fly-By-Nite U. The recognized procedure for obtaining these letters involves a long, intricate, and, at times, tedious education.

EVALUATION

Physician Heal Thyself

Although we all recognize the importance of producing a maximum number of competent and creative scientists (the Russians are getting ahead), our institutions of higher learning vary enormously in terms of the methods designed to produce such men. Interestingly enough, scientists appear to be reluctant to apply the methods of science to the evaluation of the methods of training future scientists. Since casual observation is the method of choice by most institutions, we are quite ignorant about the means of training the effective professor, the virtues of seminar versus didactic presentation of material, the value of practical training in a research setting, the usefulness of comprehensive examinations, the need for interminable essays and reports. Furthermore, few of our science teachers in university settings have had any formal training in how to teach, which causes no great concern, except, perhaps, among the student body and the odd cranky taxpayer.

Research data is available on the relative effectiveness of some educational procedures for the public or elementary school child, but there is a dearth of information on the same topic regarding the graduate student. Part of the problem lies in our difficulty of recognizing or identifying the competent and creative scientist. If a good scientist is one who produces durable packages of prized information, then, obviously, it is going to take a long time to see if his products are durable and prized. Thus, in some ways, education is like a medical treatment, the benefits of which do not become apparent perhaps for 20 or 30 years. Recall that slow-acting treatments are the most difficult to evaluate, and you will shake your head first in self-righteous indignation that university people currently rely on the crudest of methods (casual observation and bias), and then will shake it again when you realize the resources that would be required to do a comprehensive evaluation job. One consolation is that we are saving time and money by not doing such research. This advantage is perhaps slightly offset by the probability that our graduate programs are, in large part, self-satisfying tribal rituals, and colorless ones at that. Nevertheless, things are probably not as black as they seem, since there are increasing signs of interest in taking a closer look at the possible effects of some of our rituals at least.

University Commitment

Although universities are recalcitrant in evaluating their effectiveness, they are, nonetheless, supposedly in the business of evaluation. We are eager to shine the spotlight of exposure on ignorance wherever it may be—beyond the campus boundary. Thus some believe that the university has a commitment to produce scientists who will be equipped to make a contribution to society. In addition, some even believe that the university has some responsibility to protect the public from charlatans. These are noble goals, but, like most noble goals, are easier to state than to achieve.

A doctoral degree is like a magic key applied to locked portals. It opens doors to higher salaries, to sprawling homes in suburbia, to guest speaking engagements at the Chamber of Commerce, and to positions that, from the outside at least, are ones of power, dignity, and authority. By waving our doctoral key we almost automatically become respected members of the community. The assumption on the part of the public—bless them—is that the university graduate has the credentials to be regarded as an expert—a mobile source of highly prized packages of knowledge and deep thoughts. We trust and rely on keyholders to keep us healthy, to protect our country, to teach our children, and, increasingly, to run our government. It is supposedly the university's responsibility to evaluate the relative effectiveness of some of its tribal rituals, even if it can't mass-produce wise men. The university must decide, on the basis of the data at its disposal, which students to graduate, even if it can't predict whether a given student will be able to teach, or cure, or build, or administrate, or do research. There are many, many students, and if we are going to sort the good guys from the bad guys, we need decision aids to help us chunk some of the data into neat, usable bundles.

Comprehensive Exams

Most graduate schools, the final step in formal scientific training, assume that the students they accept are probably capable of mastering this final phase of training. To ensure that the student has grasped the important and current hit parade aspects of the subject matter, however, comprehensive examinations are often required. These exams are used as decision aids; they constitute a more or less formal data sample collected from all the students. Results are used to decide whether (in extreme cases) to cut the player from the team, or to make him take more courses or write extra reports, etc. Other formal data samples include the grades he receives in his courses or the ratings he receives from his supervisor. Since there's a certain elasticity in all these measures, the science novitiate can help get the stretch to his own advantage by being visible—that is in sight—so that the professors who do frequent the campus develop the impression that he is keen. Doing and handing assignments in on time scores high with some faculty, but others favor the father confessor or buddy role.

When the results of comprehensive exams conflict with well-organized biases about the student, formed on the basis of other information, creative and energetic professors usually evolve new decision rules to handle the special or embarrassing circumstances. In other words, there is no single standard yardstick for measuring graduate students. Rather, a variety of data samples—formal, informal, semantic and pragmatic— go into the hopper. The sorting, weighing, and combining procedures are done by individuals and committees. In the case of some students, there is likely to be little disagreement; like skunk oil and Chanel No. 5, they sort easily. Other students, like lovers, are rated high by some and low by others.

Before we work ourselves into another state of trembling indignation, it is perhaps wise to acknowledge that there is no simple solution to this apparent dilemma. While it is relatively simple to determine whether a student knows his professor, it is quite a different thing to determine how broadly and deeply he has sampled a knowledge area. The real question concerns the extent to which we are willing and able to invest resources in the development of shareable yardsticks for evaluation and prediction, on the one hand, and the degree to which we are willing to rely on professorial judgment, on the other. There are some who believe that professorial judgment remains the ultimate criterion and that comprehensive examination results, like the medicine man's signs, are accepted or tolerated as long as they don't come in conflict with pre-established biases. Others believe, with reasonable investment of research resources, that we can develop useful evaluative and predictive yardsticks, and therefore not only help protect the students from the capriciousness of professorial tastes, but also provide the basis on which to evaluate the usefulness of some of our training rituals. It is difficult to argue with the latter proposal, and yet it is probably equally difficult to locate university departments that are attempting to follow this course of action.

Research Tools and Rituals

If you do graduate work in the behavioral sciences, you will probably be required to become perhaps too familiar with the statistical and research rituals and the more or less esoteric measuring devices particularly suited to the relatively protected environment of the laboratory situation. For two reasons you will likely not receive much training in research strategies and procedures suited to research outside the laboratory: (1) there has been little relevant theory and methodology developed and published; and (2) the people who teach research design and statistics are usually laboratory-oriented researchers who are not very familiar with methods other than those particularly suited to laboratory settings. It is our view that the researcher who works outside the laboratory should, if anything, be more sophisticated about the principles of measurement, research design, statistics, and rules of evidence than the laboratory researcher. Note that we are talking about principles of research design, not about rituals. Too often courses in research design are indoctrination sessions covering some popular research design rituals. Though more difficult to teach, it would be preferable to provide the student with an appreciation of the main rules of evidence used in science. Hopefully, the student would then be able to select research tools to fit his research interests, rather than looking for a research problem that will fit his research tools. Give a boy a hammer and he'll go about hammering everything in sight. Give a student an electron microscope and he'll bombard everything in sight with electrons. Give him a computer and a complicated program and everything in sight will get sorted and wrapped to fit that program, whether it is appropriate or not.

THE THESIS

The major hurdle for the student in science seeking the Ph.D. or D.Sc. degree is the preparation of an elaborate report concerning the experiments he has conducted, the background of work on the problem in question, and some theoretical formulation which includes (explains or predicts) both his own, as well as earlier, work. This report, more commonly known as the doctoral dissertation or thesis, occupies a major portion of the student's energies for two years or more. It engages most of his thoughts, his dreams, and his time. It becomes the reason for his existence.

Because the thesis is an essential requirement for becoming a scientist, because the thesis requires such a large portion of the individual's life span and resources, and because it occasions so much anxiety on the part of most students, it seems worthwhile to discuss in some detail a few of the misconceptions in the practice and art of "thesising."

Thesis Myths

As with other institutions in our society, some myths have grown up around thesising which, although they may have been true in the past, appear to have little factual basis now. These myths are often the source of much distress and can even be debilitating to the thesis author.

Significant Contributions. Most university regulations regard- . ing the doctoral dissertation contain a statement to the effect that a thesis must constitute a significant contribution to knowledge. That this would be desirable, there is no doubt or disagreement. But what is a contribution to knowledge? Better yet, what is a significant contribution to knowledge? We encounter the same problem here that we had in defining what constitutes a good scientist, one who produces highly-prized and durable information. We could set up an ordinal scale, but we would find ourselves involved in acrimonious debates when we tried to get large numbers of people to endorse our definition of the categories.

Who is qualified to make such judgments? The question is essentially unanswerable because the durability of a piece of work is usually not evident for many years, as was the case with Copernicus, Semmelweis, and Mendeleev. In our view, a thesis is judged on whether the data collection and analysis procedures and rituals, and the conclusions are acceptable to a particular jury of professors—one of whom is usually from a different university. As in a beauty contest, judges may agree, but for very different reasons. We are proposing that group pragmatics play a large role in deciding whether a thesis is acceptable or not. The procedure of including an outside examiner is a commendable attempt to reduce the effect of in-group pragmatics. There is nothing particularly wrong with this procedure. It is a relatively inexpensive way of making decisions on an extremely complex issue. There may be readers who believe that theses are rated by small group semantics (by the application of a set of objective standards which can be applied by independent observers). In fact, this would make an intriguing doctoral dissertation, namely, to attempt to determine the variables and the weightings used in evaluating doctoral dissertations within the behavioral sciences. Blind ratings by a large sample of judges of theses that, unknown to them, have already been passed would provide an interesting starting point, with or without any attempt to manipulate set variables.

Standards. Another initial myth which is seriously believed by a large number of students is that the university and its professors have high, uniform, and inflexible standards against which to compare an individual's performance. Hopefully, it becomes apparent that the university is not a computer-controlled, high-class widget factory, but rather that it is a human, decision-making system, with professors and students coming in all shapes and sizes, physically and mentally, dancing to different rhythms, worshipping different idols, and at times doing what they must and calling it by the best name they know.

Some departments appear to see the graduate training program in terms of an apprenticeship, while others perceive it in terms of a steeplechase or maze-running operation. The apprenticeship model relies heavily on decentralization of all but a very few aspects of the training. It is based on the assumption that students differ markedly and the faculty differ markedly, and so centralization and standardization is bound to force the majority of students and faculty into attempting to be what they are not, into ritualistic posturing. The apprenticeship model notes that while these differences have been ignored in undergraduate work, perhaps we should acknowledge them at the graduate level, and maybe even capitalize on them, rather than seeing them as administrative inconveniences. This model, however, also assumes that grad students and faculty have the necessary brains and energy to benefit from the quite remarkable opportunity afforded them. It is upon attempting to swallow either one or both of these assumptions that the supporters of the steeplechase approach can be heard to gag—and with some justification. There are graduate students and faculty who, when faced with freedom plus time, panic and ride off in all directions at once, or sink into a state of suspended animation, or simply have one hell of a good nonacademic time. It is for this subgroup that the steeplechase model was probably designed. Once again we face the dilemma—that no single model fits all people and all circumstances. Nevertheless, we prefer the apprenticeship model, with efforts being made, prior to admission, to identify any major work allergies and, in addition, with a two-year thesis completion deadline, allowing one six-month extension, if agreed to by a hard-nosed departmental appeal board.

Failing. Even when the steeplechase model appears to prevail, the probability of failure is essentially mythical; nevertheless, it remains a real specter hanging over the heads of many students. All people dislike the feeling of having failed or having been defeated in an endeavor, but this is particularly true of the student for whom the evidence of failing is so clear-cut. You pass a high school exam or you fail, you pass your year or you fail, you pass your driving test or your swimming test or you fail. This history leaves the Ph.D. student with an ingrained belief in the possibility of failing graduate courses or, even worse, having his thesis rejected. All through the Ph.D. years embellished rumors of previously-failed students and the reason for their failure are discussed, and there is probably a feeling in the heart of most students that the next thesis to be failed could well be his. Although it is true that the odd Ph.D. candidate is failed, the excessive concern over this possibility is quite unrealistic. The Ph.D. oral, in the vast majority of cases, is essentially a formality. The exigencies of the situation are such that the university and the department doesn't want to fail its Ph.D. candidates. Thus, once a student has been admitted into the Ph.D. program, the problem of failing is remote. There is, however, a major danger; becoming a Ph.D. dropout, known as the "all but" (meaning that the individual has satisfied all the requirements for his degree except his thesis). Whether we like it or not, it is lack of focused work, rather than lousy work, that poses the real threat to the Ph.D. candidate. The large investment made by the university in each student accepted into the program is such that it can't afford to fail many graduate students. There are interdepartmental politics to consider as well. And also there is inertia in the system. University departments are slow, ponderous beasts, and this inertia favors the student once he is accepted into the program.

Prior to acceptance, if possible, but certainly immediately on arrival, a new student should avail himself of the student grapevine. This is a veritable storehouse of information and nonsense about the faculty, but, by consulting several of the senior grad students independently, very valuable information can be obtained. The new student is negligent and naive if he doesn't make a concerted effort to determine which are the faculty members with whom he can best work and learn. Some faculty are doing highly-focused research and are looking for grad students to work with them under relatively close direction and supervision. Other faculty provide general and periodic supervision, while still others provide virtually no supervision unless it is requested or perhaps even pried out of them. Some faculty get their students through in record time, while other faculty seem reluctant to have their students finish. The new student needs this information, or the myth of the standardized faculty member will prevail until the student learns the truth, often at the cost of a wasted year. If you are a highly independent, self-regulating student, you require a very different kind of thesis supervisor from the student who works and learns best under close supervision.

A Few More. There are other myths that should be noted in passing. One is the myth that someone actually knows what constitutes a Ph.D. thesis, or that there is a yardstick available in the departmental vaults that can be pulled out and applied to each thesis and so determine how close it is to making the mark. Just as a camel is a horse that has been designed by a committee, so each Ph.D. thesis is a new beastie designed by a committee in which you, the student, must play a key role, recognizing that many of the decisions must, of necessity, be arbitrary. These decisions usually constitute some kind of compromise of custom, opinion, and bias, drawn from you, your supervisor, and active thesis committee members. Once the thesis is ready for the oral examination, another myth brings undue anxiety to the student's heart. This is the myth that all members of the examining committee will have read your thesis in great detail and with complete understanding. In our opinion, in the vast majority of cases the person who produces the thesis knows much more about its limitations than anyone else. Members of the examining committee will range all the way from those who have done a conscientious job in attempting to read and understand the thesis, to those that just didn't get around to looking at it until just prior to entering the oral examination. Some of the examiners will be there to examine you, some will be there to demonstrate their own erudition, and some will do their best to indicate that they have read the thesis by discussing bits of material drawn from random points in the document.

Finally, perhaps the most deceptive myth of all is that the Ph.D. represents the last hurdle in some kind of knowledge race. A student who has just cleared the jump should enjoy this illusion while it lasts. The science game now shifts to new ground with new rules. Your rating first depends on getting some publications out; then, once you've demonstrated that you can publish, your rating depends on whether you are publishing in respectable journals; then your rating depends on whether you have a good book out; and then. . . . Well, as you can see, it's never dull, and it's never-ending.

chapter 13

Big Business

THE LARGE AND SHINY

Much of science and research is now Big Business: Big Budgets, Big Buildings, Big Staff, Big Equipment, Big Publications, Big Delays in Communication, Big Resistance to Change, Big Production for Production's Sake. Soon there will be a Broadway play titled, *What Makes White Coat Run*. Those of you who crave a quiet contemplative life in a small laboratory tucked safely away in a silent corner of a gracious and secluded ivy-covered hall can, if you find such a place, make a mint by renting it out weekends as a treatment center for the research runners who once dreamed of such a research haven, but are now addicted to the BIG.

Don't misunderstand, there are buckets of extremely important behavioral science research that don't require the BIG. In fact, a good case can be made that too many social and behavioral scientists have drifted into a research style that appears to demand big budgets for equipment, space, staff, and complex data processing. But the fact remains that in behavioral research the surface has only been scratched, so that an infinite variety of research strategies is appropriate. Therefore, you can avoid the BIG if you are the type of person who doesn't panic when most of your colleagues are getting large grants to join the race to purchase the manpower and paraphernalia that is in vogue, or that some forms of highly focused research require.

Let us examine briefly what we might call the gate receipts of the science game. How do we finance new labs and buy new players?

GRANT GRUBBING

Individual researchers, as well as great institutions, are becoming increasingly engulfed in the warm and, at times, crushing embrace of grant grubbing, spending, and regrubbing, ad infinitum. Millions of dollars are obtained each year as grants from governments, private foundations, and wealthy individuals for research supposedly related to defence, health, space, and extrasensory perception. It is reported that M.I.T., one of the greatest research institutions in the world, relies for 80 percent of its budget on defense and space research contracts.

Like most good things, grants bring positive and negative features with them. On the positive side they may provide the additional resources—eyes, ears, hands, brains, instruments—to enable you to gather more pieces for the puzzle, to allow you to attempt more alternative ways of putting them together, to afford you the opportunity of displaying your pieces or pictures to a larger audience. The advantages range from allowing you to buy more expensive research equipment and toys, to buying you relative independence from your employing organization, which may be a university, an industrial setting, or a government department. In order to obtain grants, you must compete with other researchers, like contractors make bids. The good bidder, or grubber, gets grants. With these funds you build your research empire, large or small. You buy secretaries, assistants, disciples, trips, ghost writers, early publication, and freedom from many of the usual chores of your parent organization—chores such as teaching, committee work, or projects from the boss, who may end up as boss in name only.

On the negative side, grant grubbing leads to some problems for the individual researcher and for his employing institutions. Since grants enable the researcher to hire help, he often becomes further and further removed from his data and its collection and manufacture, from an awareness of the quality of the material that went into it, and from an appreciation of the fragility or perishability of the results. As he moves in the direction of becoming a research manager or coordinator, he must rely more and more on the skill and integrity of his staff. He wants them to find pieces for his puzzle. He is overjoyed when they produce a piece that fits, regardless, perhaps, or how it was found or manufactured; the less he knows about that, the easier it is for him to accept the piece. When the pieces don't fit, he is tempted to question the adequacy of his staff before he considers the adequacy of his theory.

A Painful Dilemma

The very successful grantee faces a painful dilemma. He needs relatively senior people to help him, to act as research foremen, to supervise the technicians and graduate students. However, senior people have ideas of their own about which pieces of the puzzle have high priority and where they should be found. In other words, the more senior the person, the more independence he wants. Now the successful grant grubber moves more and more into the role of the team owner or administrator—he raises the money, but others direct the game strategy simply because he's too far removed from the playing field. The moral is if you want to keep close to your data, you have to restrict the size of your operation so that you are directly involved in all the steps. You farm out the work only after you have done a piece of it, and you work directly with the technicians and students. The further away you get from the raw data, the easier it is to fool yourself about the adequacy of the puzzle pieces and the puzzle picture.

The dilemma facing the grantee, then, is whether to hire chiefs, who want a piece of the action, or Indians, who will require time-consuming and detailed supervision. The grant system can create a problem for the university or government that wants to increase its budget through grants. Such institutions lose control, to varying degrees, over the successful grant-grubbing employee. Other staff members become disgruntled at having to concentrate on institutional chores while the grantee appears to wallow in new staff, new equipment, and press releases.

Thus, like any other resources, the benefits of grants depend upon how they are managed. Grants can and do provide many individuals and institutions with increased intelligence and more effective technology. Some individuals stagnate under the grant system and tell tales of the good old days "when I had my grants." Others become more and more proficient at predicting the grant market, and more skillful at writing payoff grant applications, but at the same time they become further and further removed from their research as they are forced to rely increasingly on senior graduate students and staff to run the business.

A Dilemma for the Graduate Student

Research grants create a dilemma of particular relevance to students. The professor must, on the one hand, honor his grant commitments by getting experiments of a given type done and published, hopefully with results that make it probable that his grants will be renewed, while on the other hand, he has an educational responsibility to the students working with him who are usually supported by his grant funds. His responsibility is to allow the student increasing degrees of freedom in being critical of experimental procedures, and in being critical of the way the professor interprets his results. He should give them the opportunity to explore ideas at odds with his own, to design and carry out experiments which may threaten, or, at least, fail to support his own theoretical position. The professor faces the dilemma of whether to select and support students who will do as they are told and so produce the things he needs to service and maintain his grants, or whether to work with bright and independent students who may fail to devote energy to his grant projects or who may hack away at some of his favorite biases.

One answer to this dilemma is to allow the grant-holding professor to hire technicians to service his grants, rather than to have him attempt to combine the teaching and grant-servicing function. The use of technicians is becoming increasingly common, particularly for senior professors. Before a general solution is found, however, university administrators and faculty must be prepared to face the problem more squarely than they have in the past.

There is competition for good graduate students. Financial support, plus professors with reputations, are the two major lures used to attract graduate students. At present, grant funds provide at least half of the financial support for graduate students, and graduate students provide the lion's share of the research grant labor. In most instances these students select thesis topics related to a given professor's grant research. By so doing, the students get access to equipment, supervision, and funds that would otherwise be hard to come by. If one student is working on a thesis close to the professor's heart and academic advancement, while another is working on a project of less concern, you can predict which student is going to have easier access to short-supply commodities like time, attention, concentration, money, equipment.

Until there is a change in the manner of financing graduate students, the problem will remain a major one, particularly for graduate students with an independent bent. Several years ago the graduate student was considered to be, among other things, a source of cheap slave labor to work on professors' research projects. Now graduate students are becoming relatively expensive and truculent labor. The students are becoming increasingly restive about doing what they see as menial work, and the professors are becoming dissatisfied with the complaints of the one-time well-behaved serfs, hence the move to hire technicians. If the trend continues, it will reduce the amount of financial aid available to graduate students. However, to complicate matters, certain granting agencies favor the use of grant funds for employing grad students and frown on, or veto, the use of grant funds for technicians. Professors relying on such funds are tempted to support the steady, reliable, good boy grad student, rather than the independent, questioning, inventive, bad boy graduate student.

There will be those who will be aghast that we should suggest that the professor faces a serious conflict of interests between his obligations of servicing the grant, or, in general, advancing his own career, and his obligations of providing his graduate students with an atmosphere conducive to critical thought and independent research. Our critics will reply that the professor's first obligation is toward his students and toward the unbridled search for durable information, and of course they are correct. There is a vast difference, however, between what should be and what is. Furthermore, it is not simply a question of intention. As every golfer can tell us, it is one thing to know what you are supposed to do, and quite another to do it effectively.

THE GRANT GIVERS

Evaluating grant applications, like judging baby contests and betting on ponies, is not a scientific activity. Grant giving is a by-guess and by-gosh affair. In most instances some kind of committee process is involved. If you have served on committees, then you have firsthand experience of the strengths and weaknesses of this form of decision making. On the negative side, you know that the intelligence of a committee probably never approaches that of its wisest member. On the positive side, you know that it is more difficult to bribe, or coerce, or cajole a whole committee than it is an individual. Like individuals, committees show favoritism, but at least the favoritism of the committee aims to keep each committee member and his constituents happy, a condition which results in a much wider funneling of the grant goodies than is likely from an individual.

While the system of grant dispersal is far from perfect, in our experience the members of grant evaluation committees are sincere men who are occasionally aware of how crude their yardsticks are for measuring the adequacy of grant applications. These men work earnestly to separate the excellent applications from the adequate and the inadequate (using an ordinal scale) according to the biases or conventional wisdom of the particular constellation of people making up the selection committee. The problems facing them are similar in some respects to those facing an investment club or a group deciding how to make bets at the track. They must decide what the chances are of a payoff, what the odds are, how much they should bet on their own horses and the horses of their trusted colleagues, how much they should go to the long shots, and, if any, which long shot. Or they wonder if they should stick with the old reliables who at least sometimes finish in the money, or if they should stake some on the young unknowns now pawing and blowing for a piece of the action.

But the horse race analogy is inadequate because the research race takes years and the finish line is vague and shifting, so that it is difficult to tell who pays off. Making book on grant giving would be a tough job. When there is no ideal yardstick, however, we manufacture make-do ones. Some of the yardsticks include: (a) how many publications by the applicant, (b) which journals he publishes in (different committees have different hit parade journals), (c) which "clubs" he belongs to, that is, whether or not he speaks the same theory, statistical, and apparatus language as the majority of the grant committee members, or, failing this, (d) whether he talks a language which committee members aspire to talk, perhaps a fancier mathematical language than their own, or whether he proposes to use an awe-inspiring mix of physical science hardware.

Like other clubs or committees, grant screening groups perpetuate themselves by reappointing each other or their methodological or conceptual kinfolk. Before clucking ourselves into indignant ecstasy over this kind of inbreeding, we should perhaps recognize that this is nothing more or less than the rest of us do when we are faced with decisions about staffing any activity in which we have a vested interest. We too appoint people we trust, people who speak our language and have our kind of values. At times these men recognize that most of our cherished truths of today will, in 100 years, be recognized as the myths of our primitive little tribe. But since the time is now, we make the best sense we can out of the multitude of alternatives buzzing around us, and we wrap the beliefs in truth cloth and call it science, or knowledge, or wisdom, or a good and fair decision.

There are obvious ways to reduce the amount of inbreeding in the grant giving process. All that is required is a mechanism to insure that some money goes to some of the heathen applications, to the non-establishment members. For example, we might draw at random from the rejected applications and give support to a certain percentage of them. Einstein was considered a heathen; many innovators are.

It is an irony that although, in calm moments, grant committee members recognize the danger of inbreeding, they nevertheless refrain from instituting simple and mildly corrective devices such as that proposed above. Perhaps such a procedure would be too constant a reminder of the temporary and provincial nature of our wisdom and of grant screening rituals.

five

Pattern
and Noise

It is proposed that the world to the newborn infant appears as a buzzing confusion. With experience he discovers order, or manufactures order, or imposes order upon this world filled with noise and shifting visual patterns. Similarly, adults, when confronted with new situations, experience a buzzing confusion or a lack of pattern. They, too, in time discover order and pattern, or impose it on the unfamiliar city, language, customs, or equipment. Recall the confusion you felt the first time you arrived in a metropolis or looked under the hood of a car, or saw the insides of a TV set—random visual stimulation with few clear-cut patterns emerging of what leads to what. Slowly patterns emerge with the help of signposts, instructions, and experience.

We gradually develop simple maps inside our heads. The more complex the situation, the more we must rely on these oversimplified maps or theories of what leads to what. Such theories must be oversimplified because we can't begin to attend to or remember the world and its multitude of tiny parts.

Not only are theories or models useful means of summarizing and bringing order into what has happened or what is happening, but they also help us make predictions that allow us to approach the future with some degree of confidence.

In this section we discuss how man, citizen and scientist, approaches the problem of handling information overload and uncertainty in his noisy world through the use of necessarily simple maps, models, or theories.

chapter 14

Theories, Maps, and Models

A theory, or a map, or a model is an abstraction, and abstractions involve representing the part of the world under examination by a model of similar but simpler structure. The important point is that theories are simpler than the data they are designed to represent. Theories are built (1) by squeezing some parts of experience together—all negroes, or all smokers, or all Southern Democrats—and (2) by ignoring or omitting some information, such as the differences that exist between negroes, or between different smokers, or between different Southern Democrats.

WHY BOTHER WITH THEORIES?

No Other Choice

We propose that the most important reason for bothering with theories is that we have no alternative. We would have an alternative if we were like some wondrous computer that (1) saw, heard, and felt everything, (2) had a massive and unlimited memory file so we could store each bit of information separately and permanently, and (3) could draw at will from our memory file each bit of information and experience for examination. Obviously, we are not like this. We listen, look, and feel selectively. We also forget, condense, and distort the information we do have. We tie bits of information together in our heads that are not necessarily tied together in the world around us. We are, in fact, information screening, condensing, and relating machines. The end results of this process are theories or mental maps of what goes together and what leads to what. If it were otherwise, we would go crazy from the great avalanche of detailed bits of information and experience to which we are exposed both from the outside and from the inside.

You will recall that when the psychiatrist is faced with too much information or too many alternative suspects he develops biases or theories about what to look at and about what leads to what. Some psychiatrists focus on biochemical information and some on early training. The important point is that by focusing on certain types of information, by developing biases, and through forgetting, we reduce the avalanche of information and experience to manageable size.

Like the psychiatrist, the scientist and the citizen increasingly face information and opinion overload. Like the psychiatrist, we too must develop simplified pictures, stories, models, or theories of what goes together and what leads to what.

Through experience, training, and bias we group together certain people, objects, or events, at the same time ignoring many differences. We may group all negroes together. Even if we do it on the basis of color, we must ignore the fact that many non-negroes are as dark as many negroes. We may go beyond simple classification and, on the basis of selective experience, bias, or training, conclude that negroes and low intelligence or laziness go together without considering the question of how many exceptions there are to this theory or picture. Others may say that low intelligence and laziness are strongly tied to early environment, ignoring the role which inheritance may play. We develop theories about women, men, communists, Democrats, Jews, WASPS, and Irishmen, the rich, the foreign, teachers, electrons, liquor, and LSD. To the extent that we omit critical information, the results of our oversight or error come home to roost, causing us varying degrees of trouble. To the extent we can screen out information contrary to our theory or bias, we continue blithely on our way.

So we propose that the first reason for bothering with theories and maps, those simplified pictures or stories of our world, is that with the information-processing equipment and ability that is built into us, we have no other choice.

Simplify Decision Making

Another reason for bothering with theories, maps, hypotheses, or hunches, is that they simplify our decision making. This point is closely tied to the fact that we have no choice but to develop and act on theory because of the nature of the beast (who distorts, omits, rejects, and selectively receives and combines information) and the nature of the world around us (with the fantastic amount of information bombarding us each second).

All of us base many of our day-to-day decisions on hunches, points of view, biases, or theories that we have developed over the course of time. Some of us have the ability to formulate a point of view very quickly, particularly if we have little information about the issue. It is much easier to develop a theory if you are relatively ignorant about the area in question. As one national leader replied to a reporter, "I would have a ready solution if I didn't know so much about it."

In any case, there is no doubt that theories do simplify decision making by providing us with decision guidelines. These guidelines are generally in the nature of a reduction in the amount and kind of information we consider, and also in the number of possible alternative decisions we might formulate. In other words, the thousand and one potential decisions we might make on any question are reduced to one, two, or three merely because we have a theory which channels our thinking along certain lines and not others.

Theories Are Fascinating

A third reason for bothering with theories is that, for many people, theories are exciting to develop and discuss, regardless of whether the theory concerns why Cathy and Bill split up or why the stock market took an unpredicted and painful downward trend just after we made our purchases. Most of us, when faced with an unexpected event, have a strong desire to explain what caused the event or how it occurred. It's like a puzzle; we worry away at it until we've got some explanation that satisfies us. It's particularly gratifying if our theory is some dramatic or startling recombination and others can be swayed to accept our new interpretation.

Predicting the Future

A fourth reason for bothering with theories is our need to predict the future. We need to predict what will lead to what even when we lack necessary information. The predictions may range from attempting to decide which of three job offers likely gives you the best chance of promotion and development, to predicting the future effects of continued smoking or of air pollution.

Scientist and non-scientist all daily act on predictions. We use the weatherman's predictions to help us decide what to wear and whether to leave earlier because of predicted snow or sleet. We predict that the car will start, or if this prediction, much to our dismay, is disconfirmed, we predict that the public transportation system will be operating, that the driver is sober, that the brakes are sound. We predict that the water supply is pure, the cook is clean, the judge is honest. We predict that in our absence our roommates will not die, and our apartment won't be burgled. Thus we're constantly operating on the basis of a variety of predictions of varying probability, often without recognizing it.

Rewards from the Scientific Community

Finally, theories are important for scientists because the academic and scientific communities offer a variety of rewards to people who build acceptable theories and who test them. The eminent men and those to whom the accolades of science are awarded are usually the theory builders.

In summary, we have theories, or simplified pictures of various parts of the world, because we have no choice. Second, we use them as decision making guides through a storm of information, opinion, and experience impinging on us from all directions. Third, for many, constructing stories about the past, present, and future is exciting. Fourth, theories about what is likely to happen assist us to move into the future with some confidence. Finally, the academic and scientific communities reward theory builders and testers.

In the final section of this chapter we will develop some yardsticks which can be used to help us evaluate the adequacy of a theory. Before doing that, however, we will briefly examine one way science can be viewed—as a love affair.

BUILDING MODELS

Theory building becomes a game of building models, or pictures of the world. Some models are based on how we see the world through our own eyes, other models are based on how we see the world through an ordinary microscope. There are models based on how we see the world through an electron microscope, and models based on how we see the world through a telescope, or how we see the world through the eyes of a protestant biochemist, or how the world looks through the eyes of a Catholic gynecologist. Once a researcher or a theorist has stabilized a picture or a model, it usually requires a great deal of time, data, argument, and turmoil before he changes its super-structure. In other words, if the data coming in through his own research, or the research of others, does not fit his picture or model, he is more likely to question the adequacy of the data, or the competence of the researcher than he is to question the adequacy of his model or picture.

A Love Affair

Since any experiment is open to criticism, the theorist always has a way out in rejecting unwelcome data. The rejection of a theory once accepted is like the rejection of a girl friend once loved—it takes more than a bit of negative evidence. In fact, the rest of the community can shake their collective heads in amazement at your blindness, your utter failure to recognize the glaring array of differences between your picture of the world, or the girl, and the data.

You will perhaps find it easier to understand some of the excitement and despair in the world of science if you view research as a love affair between the investigator and his project. During the initial courting stage he is open to certain kinds of information. But as he invests more time, energy, and money in the courtship, he becomes almost hostile to any information threatening the relationship. Perhaps as the data comes in, he has private moments of uneasiness which he shares with no one—as he analyzes the data he may have moments of agony—but it takes more than one or two lovers' quarrels to break up a love affair, which is just as well. If it were otherwise there would be almost no marriages, just as there would be very few, if any, worthwhile results from research. Fly-by-night relationships in almost any field yield little that is memorable or of lasting interest.

We suspect there are those who would disagree violently with this love affair model of research. They will say that the researcher must be completely dedicated to objectivity, that he is only interested in the truth. Perhaps there are researchers like that. We haven't met enough to fill a phone booth. We have, however, met many researchers who can be brutally objective about someone else's research project, or someone else's girl, but not about their own.

But you are no doubt wondering what happens after a while—does the love affair model shift to the marriage model? Yes, we believe so. After working with a project for a long time we gain some objectivity and can accept some of the limitations and restrictions of our model. Many of us will still keep our serious family quarrels to ourselves. The very senior and established researcher can afford to joke in public about the possible limitations of his model, and speak philosophically about the relativity of truth in science. But if you are wise and humane, you'll no more join him in making fun of his model than you would join him in his very personal game of making pseudo-fun of himself, his wife, or his dog. As you know, making fun of something you love or cherish is one thing, having someone else do it is quite another.

THEORY EVALUATION

We proposed earlier that we bother with theories for a variety of reasons: (1) we have no choice in that we are theory-manufacturing or information-condensing organisms; (2) theories make decision making simpler; (3) theories are fascinating; (4) theories help us make predictions about the future; and (5) theory production and testing are rewarded by the academic and scientific establishments.

Therefore, while there seems to be a variety of reasons for bothering with theories, how are we to decide on the adequacy of one theory as opposed to another? Generally speaking, a theory is useful to the extent it provides us with acceptable information in a shorthand or economical way, that assists us in making decisions and approaching our goals, or at least appears to help us avoid frequent high cost errors.

More specifically, if we are comparing one theory with another, we can use the following guidelines: (1) Which theory is the simplest to learn and use? (2) Which theory is more readily open to test? (3) Which theory provides us with sufficiently relevant and precise information at each step in our decision making in dealing with the question at hand? (4) Which theory provides us with the most unique and original information, or allows us to predict the most new facts or solutions? (5) Which theory best fits with other accepted facts on theories? (6) Which theory is internally consistent, that is, doesn't contradict itself?

Simplicity

We like simple theories because they are easy to remember and to apply. Scotsmen are stingy, Englishmen are cold, negroes are lazy, Jews are pushy, Wasps are self-righteous, spare the rod and spoil the child, GM products are better than Ford products. While these theories or condensations are simple, they are obviously very imprecise; nevertheless, some are widely held. So, right or wrong, the simplicity yardstick is one of the most important. This is particularly so when, even though the theory is wrong, we personally don't suffer from its application. If, on the other hand, the application of an imprecise theory obviously does lead to our discomfort, we become interested in examining

its relevance or its precision. Thus parents come to psychologists saying, "even though I've whaled the daylights out of him, he still misbe-
haves. In fact, he seems to be getting worse." Such people are then
ready for a more precise and more complex theory. Such theories might
state, "On some occasions some children respond better to reward than
to punishment," and for a more detailed background of this theory we
might even be prepared to invest in and study a book on child rearing.
Or, if the problem at hand is buying a new car, we may, through bitter
experience, be forced to subscribe to the theory that only on certain
years are some GM models better than Ford models, and acquaint
ourselves with the theories or condensations of such publications as
Consumer Reports. Notice, then, that increased precision is usually
purchased at the price of increased complexity in our theories. The
general rule is: The more accurate we want to be, then usually the more
complex the theory, and the more information we have to include in
reaching our conclusion.

You can usually spot a simple theory by its emphasis on one or two
bits of information: behavior depends essentially on race, or behavior
depends essentially on early training, or behavior depends essentially
on biochemical factors, or behavior depends essentially on punishment,
or behavior depends essentially on the institution that you are working
for.

While any one of these one-cylinder theories may have some validity,
the more precision we want in our solution, the more likely we will need
to combine these one-cylinder theories into multi-cylinder theories; behavior depends on intelligence, and on early training, and on genetics,
and on biochemistry, and on work institution.

So we face the problem of reaching a balance between simplicity and
precision. If we want to predict the developing behavior of a child, we
may have to combine the series of one-cylinder theories. If, however, we
merely want to predict the behavior of a particular bus driver (e.g.,
when he will come to our stop), we will usually be adequately accurate
by consulting the bus schedule published by the institution that employs
him. In most cases it would be irrelevant to worry about his intelligence,
his early training, his genetics, and his biochemistry.

Therefore it is not a question of simplicity versus complexity, it is a
question of whether a theory or condensation includes enough information to meet our needs or help us make a decision.

If we want to understand or predict complex human behavior, then
theories, of necessity, must be fairly complex because (a) people differ, (b)
they learn, (c) their behavior changes from one situation to another, and
(d) they change with age. Therefore any time you encounter a theory
about human behavior which is based on the assumption of stability,

you will realize that such a theory leaves out a great deal of information. Examples include theories about introverts and extroverts, depressives and nondepressives, laziness and activity, responsibility and irresponsibility, etc. The reason why these theories have some appeal is: (1) they are simple, and (2) there are a few people who fit them. But they leave out most people. Most people are sometimes extroverted and sometimes introverted, sometimes lazy and sometimes active, sometimes responsible and sometimes irresponsible, sometimes security-seeking and other times risk-taking. Therefore, the next time you hear a speaker describing a theory about human behavior that divides people into a few simple classifications, you can bet that he is leaving out a great deal more than he is including, and that the major appeal of his theory is on the basis of simplicity.

Testability

While simplicity and personal appeal are two very important yardsticks in theory evaluation, the yardstick of testability is one which is presumably the most important used in science.

Think, for example, how you would go about testing the theory that negroes are less intelligent than whites. It would be a relatively simple matter if we wanted to test the theory that negroes are taller than whites. We could obtain a nonelastic ruler and measure a large, random sample of negroes and a large, random sample of whites. But intelligence tests are elastic rulers—the scores depend on (a) how the tests are administered and by whom, and (b) whether the people being tested and compared have been exposed to similar educational opportunities. Perhaps eventually some researchers will develop an acceptable test of intelligence that is as nonelastic as a tape measure, or at least not as elastic as the tests we now have, but, until better tests are available, the theory is relatively immune to test.

Similarly, if we wanted to test whether GM products are better than Ford products, we need a series of nonelastic yardsticks of what we mean by better, and then we need to measure a large number of different cars manufactured by the two companies. The reasons that make some theories difficult to test or evaluate include the following: (1) lack of nonelastic yardsticks; (2) inability to agree on which of the available yardsticks to use; (3) inability to measure a large, representative sample of the population we are theorizing about; and (4) citizens and scientists who refuse to change their minds even in the face of the new information. With some theories it is difficult to agree how they should be tested, and with others we are not prepared to invest the resources necessary to test them.

Novelty

Theories which lead to surprising or novel information are highly valued. Thus Theory A raises few eyebrows, stating, "Students will get higher marks if they do one hour's study a day for ten days rather than if they do ten hours a day for one day." Eyebrows shoot high, however, if the theory states, "Students will get higher marks if they are exposed to a tape recording of lecture notes for one hour a night while they sleep, for ten nights, than by spending one hour a day for ten days studying the same material." Of course, theories that lead to novel or surprising information are not necessarily readily accepted. Unless such theories lead to an overwhelming amount of evidence, or permit a large number of researchers to test them readily and obtain the same results, these new findings may languish for many years in obscure journals until more and more evidence accumulates, or until the biases and attitudes of a sufficiently large number of the population change so that the new information becomes acceptable.

Goodness of Fit with Other Facts and Theories

As we noted at the outset, a few new facts do not change a well-established theory or lead to the acceptance of a new theory. This is not so merely because some scientists are biased, but rather because we all use familiar theories to evaluate new information—to help us in our decision making. If we gave every new bit of information or theory careful consideration, we would be overloaded with work in ten seconds. For example, drug companies put hundreds of new drugs on the market each year, and only a few can be adequately evaluated merely because of the time and effort required. Furthermore, thousands of research studies are published each year, but only relatively few are repeated by other investigators. Most investigators are busy preparing to publish their own research.

There is no simple solution to this problem other than using our own personally accepted theories, or small modifications of them, as guidelines in helping us decide what we will read and examine carefully.

Internal Consistency

Another way of classifying or evaluating theories is to assess the internal consistency of the theory. This is perhaps one of the minimum conditions a theory must fulfill if it is to be seriously considered. A theory which contradicts itself on any specific question proves embarrassing at times, even though, taken separately, each part is acceptable. Consider the following examples: (1) He who hesitates is lost, and fools rush in where wise men fear to tread; (2) "Colonel Cathcart was conceited because he was a full colonel with a combat command at the age of only thirty-six; and Colonel Cathcart was dejected because although he was already thirty-six he was still only a full colonel."[1]

There are contradictions in the above examples and with such contradictions the theory provides no overall guidelines to aid us in predicting or in making decisions. According to theory (1) we should both buy and not buy the speculative stock, and according to theory (2) we really can't predict whether Colonel Cathcart is happy or dejected. Until we have more information about the conditions surrounding lost opportunity because of hesitation, we can't use theory (1) effectively.

Scientific or philosophical theories, as a rule, do not have such gross or obvious inconsistencies; however, often inconsistencies exist and are fair game for long arguments.

Understanding and Prediction

You will no doubt have heard that the purpose of theories is to help us both understand certain parts of our world and make predictions leading to new information and new solutions.

Understanding is a difficult term to define. Perhaps one of the most common meanings of the term "understanding" is that we develop, sometimes with the help of others, a more satisfying picture of some part of the world. For example, we may ask, "What's University like?" The reply may be, "University is like high school except that no one cares if you come to class, and in most classes you don't get any comments on the work you hand in." Or you may ask, "Why is Harry so cranky?" And you may get the reply, "He had a fight with his girl friend." Typically you will respond, "Oh, I understand." You understand because you can now combine some bits of information that you already have in a new way—you now feel you have a better picture to go on. Our point is that understanding may give you a more acceptable picture about a part of the world, but, nevertheless, that picture may be quite erroneous. Unfortunately, you can walk away feeling you understand from a wide variety of different replies to the same question. It is proposed then that the term understand implies a personally-acceptable picture of what goes together or what leads to what, but need not imply an accurate picture. The accuracy is subsequently, if ever, determined by more direct personal experience or by more accurate additional data from another source.

The term "prediction," on the other hand, if stated in testable form, is

---

[1] Joseph Heller, *Catch-22* (New York: Dell, 1955), p. 192. Copyright © 1961 by Joseph Heller. Reprinted by permission of Simon & Schuster, Inc.

a more scientifically useful concept. Theories that enable us to understand are personally useful, while theories that enable us to predict are both personally and scientifically useful.

This does not mean that the term "understand" cannot be redefined to include tests of the adequacy of the information; it is merely that the term, as commonly used, does not usually imply such tests whereas the term "predict" more frequently does. Therefore, we propose that theories that predict new information should be ranked more highly than those which lead to understanding, as defined above.

We have stressed the importance of the testability of a theory. Testing a theory can be done in several ways. A theoretician may make a specific prediction growing out of his theory and may state his theory in a testable form which we then subsequently test and support or refute; or he can do an experiment growing out of his theory, and we can attempt to repeat his experiment. Replication in science is probably the foundation of testability, but, as noted earlier, we prefer the concepts of exportability or perishability of data to that of replicability.

Summarizing, we are more and more inclined to view theories as decision-making aids, perhaps not as whimsical as the toss of a coin, nor as crude as a race tout's tip, but decision aids, nonetheless. In the face of ignorance, but forced to act, we build theories—these simplified and crudely drawn maps of the past, present, and future—that give us some semblance of confidence as we race or stagger through life's great maze of alternatives. Are theories true or false? No one knows, since most theories are designed to cover mammoth areas or massive populations, and yet we can usually explore in detail only a tiny corner or a few instances. Thus whether a theory is true or false is anybody's guess, whereas everybody knows that a decision aid is worthy if it helps you make even one decision that is not immediately followed by a disaster.

chapter **15**

The
Truth Spinners

Last chapters furnish authors with the final opportunity to include or elaborate cherished biases, and also provide a safe place to play with nice, fat, controversial topics. Here we can say our piece and run.

THE DATA SYSTEM

For a moment imagine yourself, the human data processor, immersed in life—in that flow of data that reaches back into the past, forward into the future, and even in the present flows beyond your reach, in the distance, around corners, behind your back, behind someone's eyes, or words, or skin. But stop—don't visualize too well this vast data flood or you'll be swamped, lost in a whirlpool of audiovisual noise.

In the previous chapter we proposed that the main value of theories, or models, or biases is that they serve as decision aids—a way of handling the data flood. They are man's salvation from annihilation by audiovisual-kinesthetic noise. An unbiased or unpatterned view of the data stream would be a disastrous buzzing confusion.

Bias

A bias is a predisposition for one way of seeing, feeling, thinking, or behaving, over another—a tendency, mild or strong, to follow or choose one alternative instead of another. Some of our biases, our decision aids, are clear-cut ones like reflexes, habits, or neat classifications such as "all women are inferior" or "all negroes are lazy." Some biases are mildly qualified, such as "most women are inferior" or "most negroes are lazy." Still other biases or points of view are very general, such as "women think differently" or "negroes think differently."

A bias may be a highly predictable and frequent way of acting, as with a habit, or it may be a tendency to think (for example, about the purpose of life) in a certain way when the question comes up, which may be rarely. Furthermore, a biased way of thinking, or believing, or talking to oneself, does not necessarily mean that the particular bias (I am a Christian) is apparent in much of the individual's overt behavior—except perhaps for special occasions (Sunday mornings). So while some biases may be very evident in the way the person usually thinks, feels, and acts, others may be bull session or special occasion biases like philosophies of life and special table manners. Nevertheless, these latter biases are still decision aids for these special circumstances, and can be helpful in predicting or understanding behavior.

Bias provides our basis for personal and organizational navigation. We rely on bias because we must—whether the bias comes from a fabricated T.V. commercial, or a unique personal experience; from poor aim or a reflex; from naturalistic observation, case study, or control group experiment; from an expert or a parent; from a poor memory, or an active imagination; from the impulsive thrust of panic, or the gentle drift of fatigue.

The Vice Versa Dilemma

Because the data stream surrounding us consists of different kinds or classes of data, truth seekers, knowledge builders, and decision makers are faced with a great dilemma—seeing is believing, and vice versa. On the one hand, our world includes some hard, clean-cut, here and now chunks of experience, like skunks and rocks, about which nearly every human can agree regardless of politics, religion, or personal taste. Here, to a great extent, seeing is believing—the true-false approach to the world works well. On the other hand, our world also includes millions of bits and pieces of data widely scattered, often hidden, and always changing—negroes, WASPS, viruses, and women—about which we must make decisions based on tiny samples. Here, to a large extent, believing is seeing. With partial data, or with data that ebbs and flows and hides, it is difficult to apply true or false labels about which different people can agree.

Resolving the Dilemma. How are we to handle this vice versa dilemma? One way that provides some security, if only for a short time, is to forget that we usually deal with tiny samples of data and simply go ahead and nail truth labels to all our guesstimates, implicitly assuming that seeing is believing. Another way is to reserve sufficient energy for continuously updating our guesstimates as new data comes in, and as new data analysis methods are developed. But just as the previous method of treating our guesstimates as truths has the disadvantage of providing a false sense of security, so, too, this constant revision approach has a disadvantage because it requires infinite time, constant vigilance, and superhuman tolerance for ambiguity. People usually handle the dilemma by using both these approaches. On the one hand, we reserve a few high priority spots in the data stream (golf, or women, or cooking, or stocks) for vigilance, updating; and practice; on the other hand, for handling the bulk of the data stream we rely on bias, our own or someone else's. We treat guesstimates as truths, and continue to do so until a bias that fits us better comes along, or until we are forced to change because of a punch or two in the attention.

SCIENCE AND BIAS

Scientists, like citizens, rely on biases as decision aids. When they operate outside their own specialty, they typically rely on the same biases as the citizen. When they operate within their specialty, they rely on a variety of biases, some common to science, some common to their specialty, some unique to the individual scientist. Some scientists have a strong semantic bias, a bias toward collecting data—these are the empiricists, the data-oriented, data-bound scientists; others have a strong syntactic bias, a bias towards building models or general theories with relatively less concern about data; while still others have a strong bias toward proving that their pet theory or model is right, and busy themselves in selecting data that fit their theory.

Just as there are different ways of thinking about religion and politics, there are different ways of thinking about science. Some scientists who are strongly data-oriented, for example, think about science somewhat differently from scientists who are strongly oriented towards model building, while still other scientists will think about science one way during the data collection stage of their research and quite another way during the theory building or evaluating stage. At this point we will consider briefly two alternative points of view or biases about science, and, following that, develop a simple model or way of thinking about the operation of bias, whether found in scientist or citizen.

Science: Discovery or Invention?

Francis Bacon, considered by some to be the father of modern science, over three hundred years ago had this to say about nature: "Either she is free and develops in her own ordinary course, . . . or she is constrained and moulded by art and human ministry."[1] This is one of the most fascinating areas of disagreement that can be seen emerging among researchers—viewing science on the one hand as discovery, or on the other hand as invention. Such a distinction is not brand new as the above quote from Bacon indicates.

Discovery. If you think about science as a process of discovery, you can assume that nature comes in packages or pieces, like a jigsaw puzzle. The job of the scientist is to find pieces and see which ones fit together—the characteristics of the pieces determine how the puzzle turns out.

If you like the discovery model, then you can readily decide who is a good researcher. Good researchers find good pieces—pieces which are hard, stable, and have cleanly cut edges so that other researchers can recognize them, can tell one piece from another, and can examine them closely without having them fall apart. Such pieces are like the hard, clean-cut, here and now chunks of experience, or like well-controlled studies with large stable samples. Poor pieces, on the other hand, are soft, and malleable, with frayed edges, and are loosely tied together or are hard to find. When presented with such pieces, other discovery researchers find it difficult to tell one from another, and the pieces seem to change with time, or crumble away when handled or examined closely. Some scientists who have a pet theory, however, like such soft pieces because they can be squeezed into the puzzle almost anywhere and made to fit. If the manufacturers of jigsaw puzzles put a few soft pieces into every puzzle, things would be much simpler, but to some it would be like cheating—you would never be sure what the real puzzle would look like if all the pieces had been hard and cleanly cut.

Those who accept the discovery model of science, regardless of whether they are currently working with hard or soft pieces of the puzzle, assume that it is just a matter of time until the real or true puzzle will be completed—the right answer found. Faster, more elaborate computers, a solid rocket fuel, larger samples, better personality tests,

---

[1] Fulton H. Anderson, ed., *The New Organon and Related Writings* by Francis Bacon (New York: The Liberal Arts Press, 1960), p. 273.

another Einstein, more sophisticated experimental designs, a trip to Mars, mapping the brain, a wonder drug—and nature's mysteries will be solved; the real picture will emerge.

Invention. But the jigsaw puzzle model is only one way of thinking about science. A different approach, and perhaps one that will appear strange to you, is to think of science as invention or creation, rather than as discovery. This approach is more akin to the believing is seeing side of the dilemma discussed earlier. Here you assume that the world is not necessarily divided up into neat pieces waiting to be discovered, but, rather, it is a great buzzing confusion, or a vast data stream flowing out of the past into the future and surrounding the present, reaching beyond our horizons into every nook and cranny of map and mind. Man, with his peculiar sensors and brain, imposes order on the buzzing confusion or data stream, then shouts, "Look what I found." In other words, he draws a circle around a bit of experience, he draws it with his biased sense organs and nervous system and with instruments. With this view of science, a researcher is a mold maker who presses parts of the data stream into his mold and says, "See what I've made. See, here is one way the buzzing confusion can be viewed."

It is easy to grasp the relevancy of this approach to data that are soft, distant, hidden, widely dispersed or changing, but it can apply equally well to the apparently hard, unambiguous data bits. For example, what is the real way to look at a desk? It depends on the instruments you use and the mental biases you have about what is real or what is right. From an executive's point of view, his mahogany desk is his monument; from a termite's point of view it's a great Christmas cake, from the average citizen's point of view a desk is a very solid object, from a physicist's point of view it's so full of holes it can't even stop a little old beta bundle. Thus experience and reality become relative; they depend on the nervous system, the biases built into it and acquired, and on the extensions to the nervous system that we use—telescopes, microscopes, X-rays, gamma rays, and other boundary-drawing concepts and equipment still to be invented.

With this approach, there is no true or false, no right or wrong way of ordering the world, there are a multitude of ways; there is not one reality, there are a multitude of realities.

SCIENCE AS WE SEE IT

Science, as we see it, is the process of construction or creation, rather than discovery, of pieces of the puzzle by bringing biased sensing devices—sense organs, brains, points of view, instruments—in contact with nature, and then fitting the constructed pieces together as best we can with the system of biases operating at the time and with the resources at our disposal. SCIENCE BECOMES THE GAME AND ART OF DESCRIBING A PATTERN WITHIN A SYSTEM OF SENSING AND CONCEPTUAL BIASES OR LIMITS.

We are all aware of the critical part bias plays in some areas of human discourse; politics and religion are two clear examples where the biases of others, but not our own, become patently obvious. Even in something as simple as a traffic accident witnessed by several people, the role of biased observing, remembering, and reporting is common knowledge to law enforcement and legal authorities. Disagreements arise over all facets of the accident. Because of such disagreements between observers of the same event, or between people who have different samples of widely dispersed data, viewing science simply as discovery seems an inadequate model. These disagreements suggest that man is not a simple detector, recorder, and reporter of nature's pieces. Alternatively, seeing science as creation or invention helps to account for disagreements between observers of the same event, or between researchers sampling from a great data stream. The invention model proposes that man and nature combine to make patterns—man processes experience—he spins his own truths.

Man, the Pattern Maker

Some of man's tapestries or data packages result because he starts with a homosapien nervous system or experience molding system rather than that of a frog or a martian. Some of his patterns result from a combination of homo sapien nervous system biases backed up by WASP culture molds or biases, or Southern negro molds or Eskimo molds; some of his patterns emerge from a combination of homo sapien biases, WASP biases, and hungry belly and fatigue biases; some patterns emerge from homo sapien molds, backed up by WASP culture, university training, and L.S.D. Some molds or biases he inherits from his genetic pool, some from the structuring of his early experience, some from friends, some from the press, some from ignorance.

Man gradually produces patterns, or the semblance of patterns, out of the data flood, out of the buzzing confusion. Given his sensory apparatus and experience, some of the audiovisual bits and pieces are chunked and tied together into a semantic world. For example, all those millions of bits and pieces of green and brown can be chunked together into one forest, and all those thousands of moving pink and white planes can be chunked together into one girl, and the multitude of shifting frequencies and tones can be tied together into words and sentences. These packages we apparently share in common with other homo sapiens in the data stream.

But not only do we bring to the data stream our human reflexes and our human capacity to tie together in our minds bits of experience that stand together (leaves and trees), or move together (language sounds), but also, because our place in the data stream is unique, we tie together in a unique way some bits and pieces of experience. Not only is there a forest; for one it is filled with ants and bugs, for another with tranquil glades. Not only is there a girl; but, for one, she is too thin and for another just right. Not only are there words, but behind the words one detects a cold soul, while another hears a tentative invitation to love. Which is the true picture? Who can say? It is like asking whether a cup of water drawn from a fast-moving stream is true or false. It is a tiny package of data sampled from a great stream. No one will ever use the identical cup, nor stand in exactly that spot again.

THE DATA PROCESSING MODEL

Earlier we proposed that science is a description of pattern within a system of sensing and conceptual biases (species biases, subgroup biases, personal biases). But the reader must now be feeling that we have succeeded only in transforming the buzzing confusion of audio-visual-kinesthetic noise into a buzzing confusion of biases. We need some bias in our view of biases; we need a simple model of bias as a memory or decision aid. Note that we say "*a* model"—you may prefer to invent your own.

A Five Phase Model

According to our model, patterns are created through the interaction of five bias systems.[2] (1) Bias can reside naturally in the data (rocks, skunks, explosions), or it can reside in the data artificially in that it has been prepackaged for the rest of us by a few experts, authorities, or friends, on the basis of their tiny samples (Indians are lazy, Wasps are self-righteous, women are. . . .). (2) Bias is a function of the characteristics of the sensing apparatus—the eye alone or the eye with the aid of X-ray equipment or with the aid of microscopes. (3) Bias may be found in the conceptual apparatus—the data classification and belief system, the habits of thinking, the data analysis, sorting, weighing, storing, and retrieving strategies or rituals. (4) Bias may reside in the behavioral response system. For example, the conceptual system may compute a course of action (punch the bully in the teeth) that the response

---

[2] The five proposed stages of data processing are conceptually convenient, but arbitrary—for example, raw data bias and sensing bias are considered highly interdependent.

system is unable to execute because of inadequate motor skills. (5) Bias may be due to a feedback or consequences system which is incapable of detecting negative results of a response until it is too late (smoking or talking). The model is presented schematically in the following table.

How It Works

The greater the number of alternatives there are in each stage in the data processing system, the greater the potential processing costs in terms of time, effort, processing resources, and frustration involved in sorting and selecting among alternatives. Thus the more alternatives we face, the more need there is for strong bias to protect us from decision-overload. Since the data stream is infinite, and since man's data processing resources are not, he must rely on methods which reduce drastically the number of alternatives at each stage. For example, most of us are prepared to consider all skunks alike. There are usually no alternative sensing responses but STRONG, no alternative conceptual responses but BAD, and no alternative behavior responses but RUN. With no alternatives at each stage, the data processing is fast and easy. But as soon as even one alternative exists at any stage the data processing costs and efforts rise—one pet skunk in the area plus wild ones provides you with alternative sensing, conceptual, behavioral, and feedback responses, and consequently with increased data processing costs. Thus one of the ways of reducing alternatives and thereby reducing data processing costs is to rely on reflexes and habits—when raw data triggers senses, then in these instances there is no need for complicated conceptual gyrations or response selection or shaping—all phases of the data process run off almost automatically (skunks—run; explosion—duck); all represent automatic low-cost data processing.

If the input is people instead of skunks, the possible alternatives at each stage are extremely great, and the permutations and combinations across stages are astronomical. We reduce the number of alternatives by delegating most of the data processing to individuals (experts and friends) and institutions. Delegation can be relatively complete (medicine: where we are more or less completely dependent on the doctor's data sampling, sensing, conceptual, behavioral and even feedback biases—"you're better now, you can go home"), or delegation may only involve one or two stages (raw data and sensing), as is supposedly the case in a good news story in which you decide how to analyze the particular news story and also decide on what course of action should be taken, and what the results were or will be, or you may delegate these phases as well by turning to the editorial page.

But a constant diet of prepackaged information is as dull as a constant

Table 3. BIAS IN A DATA PROCESSING SYSTEM

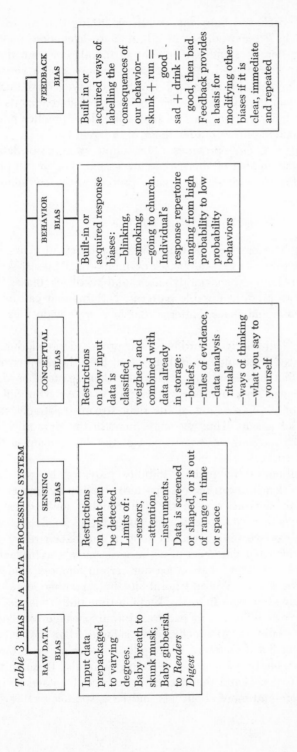

RAW DATA BIAS	SENSING BIAS	CONCEPTUAL BIAS	BEHAVIOR BIAS	FEEDBACK BIAS
Input data prepackaged to varying degrees. Baby breath to skunk musk; Baby gibberish to *Readers Digest*	Restrictions on what can be detected. Limits of: —sensors, —attention, —instruments. Data is screened or shaped, or is out of range in time or space	Restrictions on how input data is classified, weighed, and combined with data already in storage: —beliefs, —rules of evidence, —data analysis rituals —ways of thinking —what you say to yourself	Built-in or acquired response biases: —blinking, —smoking, —going to church. Individual's response repertoire ranging from high probability to low probability behaviors	Built in or acquired ways of labelling the consequences of our behavior— skunk + run = good - sad + drink = good, then bad. Feedback provides a basis for modifying other biases if it is clear, immediate and repeated

diet of prepackaged baby food with everything that might upset us or make us chew removed. So, in addition to delegation, another way of alleviating boredom and yet preventing inundation from the data flood is to focus your limited data processing resources on a very few areas—golf, or blondes, or studying the courting behavior of the confused fruit fly. Here you go out of your way to sample different parts of the data stream with a variety of sensors, play with ideas, practice new responses, and are open to or seek out feedback of different kinds.

The following table provides a few simple examples of data processing in which the number of alternatives are varied, though still pitifully small compared to the number we usually process every few minutes.

Pattern Durability and Change

The products of man, citizen and scientist, are the patterns created within a system of biases or limits—data biases, sensing biases, conceptual biases, behavioral biases, and feedback biases. The aim of science is to produce durable patterns, patterns that can be generalized through time and space, patterns that hold over wide areas of the data stream.

Durability. According to our proposed model, durable patterns are produced by stable biases, biases that limit the number of alternative patterns that can emerge. Presumably, then, pattern or data package durability can be produced in a variety of ways. Durability can be a function of biases existing in the data stream—relatively stable bits of flotsam and jetsam. Thus we may increase the durability of our data package by focusing on a small part of the data stream—a tiny segment in time or space.

In addition to the pattern stability provided by our homo sapien nervous system, we can increase pattern stability by further standardization of our nervous system through training and educational rituals (standardize our data collection or observing methods and instruments). Not only can we pick a certain spot in the data stream, but we can further limit what data gets to us because our observing and measuring instruments are filters that screen out certain temporal and spatial parts of the data stream. We add durability by screening some data out, by letting only some data in, or by keeping time out. Thus one of our ways of adding durability to our patterns, to our packages of data, is to take tiny static samples, to pluck a tiny sample from the data stream, to restrict its context or freeze it—literally, or on film, or in our notebooks, or in our recording apparatus, or in our journals. To some this is precision and is considered the essence of science, to others it is a game of saying more and more about less and less. Another strategy is to take a

Table 4. SIMPLE DATA PROCESSING

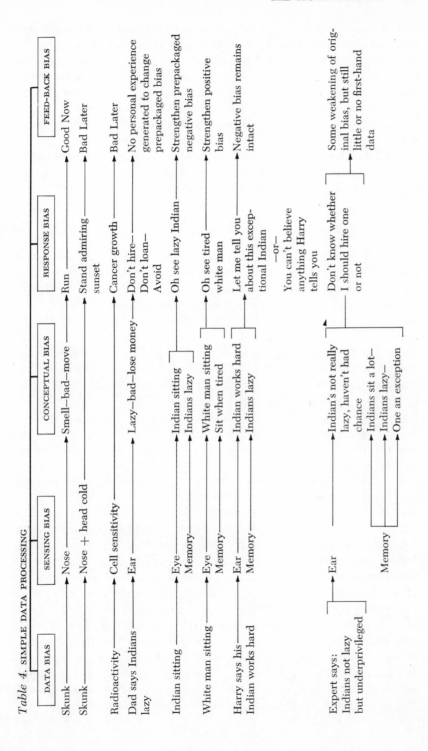

DATA BIAS	SENSING BIAS	CONCEPTUAL BIAS	RESPONSE BIAS	FEED-BACK BIAS
Skunk	Nose	Smell—bad—move	Run	Good Now
Skunk	Nose + head cold		Stand admiring sunset	Bad Later
Radioactivity	Cell sensitivity		Cancer growth	Bad Later
Dad says Indians lazy	Ear	Lazy—bad—lose money	Don't hire— Don't loan— Avoid	No personal experience generated to change prepackaged bias
Indian sitting	Eye / Memory	Indian sitting / Indians lazy	Oh see lazy Indian	Strengthen prepackaged negative bias
White man sitting	Eye / Memory	White man sitting / Sit when tired	Oh see tired white man	Strengthen positive bias
Harry says his Indian works hard	Ear / Memory	Indian works hard / Indians lazy	Let me tell you about this exceptional Indian —or— You can't believe anything Harry tells you	Negative bias remains intact
Expert says: Indians not lazy but underprivileged	Ear / Memory	Indian's not really lazy, haven't had chance / Indians sit a lot— / Indians lazy— / One an exception	Don't know whether I should hire one or not	Some weakening of original bias, but still little or no first-hand data

Table 4. SIMPLE DATA PROCESSING (cont'd)

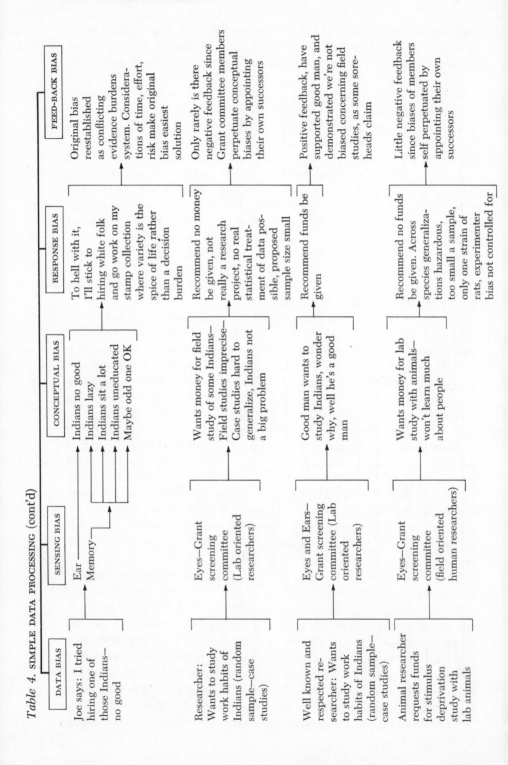

DATA BIAS	SENSING BIAS	CONCEPTUAL BIAS	RESPONSE BIAS	FEED-BACK BIAS
Joe says: I tried hiring one of those Indians—no good	Ear / Memory	Indians no good / Indians lazy / Indians sit a lot / Indians uneducated / Maybe odd one OK	To hell with it, I'll stick to hiring white folk and go work on my stamp collection where variety is the spice of life rather than a decision burden	Original bias reestablished as conflicting evidence burdens system. Considerations of time, effort, risk make original bias easiest solution
Researcher: Wants to study work habits of Indians (random sample—case studies)	Eyes—Grant screening committee (Lab oriented researchers)	Wants money for field study of some Indians—Field studies imprecise—Case studies hard to generalize, Indians not a big problem	Recommend no money be given, not really a research project, no real statistical treatment of data possible, proposed sample size small	Only rarely is there negative feedback since Grant committee members perpetuate conceptual biases by appointing their own successors
Well known and respected researcher: Wants to study work habits of Indians (random sample—case studies)	Eyes and Ears—Grant screening committee (Lab oriented researchers)	Good man wants to study Indians, wonder why, well he's a good man	Recommend funds be given	Positive feedback, have supported good man, and demonstrated we're not biased concerning field studies, as some sore-heads claim
Animal researcher requests funds for stimulus deprivation study with lab animals	Eyes—Grant screening committee (field oriented human researchers)	Wants money for lab study with animals—won't learn much about people	Recommend no funds be given. Across species generalizations hazardous, too small a sample, only one strain of rats, experimenter bias not controlled for	Little negative feedback since biases of members self perpetuated by appointing their own successors

wide-angle view of the data stream, many bits and pieces changing in space or time, and to look for trends. To some this is the essence of science, to others it is a game of saying less and less about more and more.

We can add further durability to our data packages by standardizing our conceptual apparatus (our thinking, our logic, our data analysis procedures) by only accepting those statements which pass a more or less arbitrary statistical test, or the judgment of one editor, or the judgment of two reviewers, or of a thesis or grant committee.

Further, we can increase stability of our patterns by standardizing some of our overt behavior, by developing uniform, well-established habits or ways of responding (if Indian—don't hire; if pupil—give I.Q. test; if research data—do analysis of variance; if research report—read at Eastern Psychological Association meeting; if research manuscript—submit to journal A and, if rejected, submit to journal Y; if criticized after publication—use one of three standard replies to criticism).

Finally, we can increase stability by avoiding negative feedback. This can be done by working in low or vague feedback areas (like education where yardsticks are vague) or in fields in which institutional protection from negative feedback exists (high cost of appeal in courts, time and effort to get negative feedback into the decision systems of some professional groups, e.g., medicine, or through to key decision makers in government or industry, as most negative feedback is screened out along the way). In science, for example, there is little or no provision for the publication of negative results.

Change. We have proposed that the durability of the products of scientist as well as citizen are a function of uniform biases existing at different stages in data processing. Thus many alternatives are never considered, and the creation of unique patterns is rare. We have established that consideration of alternatives involves phenomenal resources which may be unavailable to the citizen. To some extent, however, we assign extra data processing resources to the scientist so that he is able to analyze more alternatives and perhaps thereby produce a new pattern.

By providing our researchers with extra resources to explore the data stream repeatedly in spots we haven't been, with new sensing devices and strange ways of thinking, we insure some flexibility and change in our biases and patterns. At times our researchers reward us for our generous support by sending one of their number on a trip into outer space, or by amplifying our feedback systems to point out that one of our simple pleasures is probably leading to lung cancer, an economic depression, or bad breath. And while most of us ignore these dire warnings, other researchers work to minimize the risks or to develop cures.

Some durability in the patterns produced by scientists is guaranteed, because once a data package gets by the number of sieves or biases set up by the scientific establishment it is sometimes as laborious to reverse the decision as it is to reverse a Supreme Court ruling. In both instances massive resources are required.

For as long as we use crude yardsticks, we fail to detect differences and we miss change; as long as we restrict who or what gets into our sample, we miss differences; as long as we observe our sample only once or only briefly, we miss change—but notice that in the process of missing change, we capture durability. And durability remains captive until the ever-crude yardstick becomes a little more precise, until we take another sample, until we observe the old sample again, and until this new data package is shaped to fit the data processing system of the particular scientific specialty.

Thus, in science, whether the durability exists at the data stream level or at the sensing level, or at the conceptual level, or because negative results were not seen is irrelevant. It has received high-cost data processing, and it will take high-cost data reprocessing to change it. And surely this is the name of the game—to ensure that we provide curious and active people with large data processing resources and wonder at the new worlds they produce, some way out there, some right in our own back yards.

THE GREAT ENDEAVOR

In summary, the model of man we propose, scientist or citizen, is that of an astonishing pattern maker, a truth spinner, immersed in a great data stream. Here he tries to produce durable packages of data within his system of sensory and conceptual biases—produces enough pattern so he can make his pronouncements, wiggle his steering wheel, cure cancer, and perhaps even blow up our data stream.

Science, like anything of consequence, is a tangle of Gordian knots, and while the topic is inexhaustible, readers and editors are not, so an end must be found. Although it is now the time to close, we are far from closure. We like to feel that our failure to deliver a neat and tidy story is not wholly due to our own ignorance and sloth. In addition, we believe that writing about any worthy topic is like fighting bees—defend yourself on one quarter and you expose yourself to instant attack on another. To have produced a tidier tale would have meant artificially shielding ourselves, and you, from some of the more stimulating bees.

Finally, if some of our readers conclude we have treated science in a cavalier manner, then we have failed to convey our deep affection for and commitment to sciencing—man's noblest game and, in our view, his finest endeavor.

Selected Readings

Anderson, B. F., *The Psychology Experiment.* Belmont, Cal.: Brooks/Cole Publishing Company, 1966.

Anderson, F. G., ed., *The New Organon and Related Writings,* by Francis Bacon. New York: The Liberal Arts Press, 1960.

Asimov, I., *The Intelligent Man's Guide to Science.* New York: Basic Books, Inc., Publishers, 1960.

Bachrach, A. J., *Psychological Research: An Introduction.* New York: Random House, Inc., 1962.

Baird, D. C., *Experimentation.* Englewood Cliffs, N.J.: Prentice-Hall, Inc., 1962.

Barzun, J., *Science: The Glorious Entertainment.* Toronto: University of Toronto Press, 1964.

Beveridge, W. I. B., *The Art of Scientific Investigation.* New York: W. W. Norton & Company, Inc., 1957.

Blalock, H. M., Jr. and A. B. Blalock, *Methodology in Social Research.* New York: McGraw-Hill Book Company, 1968.

Braybrooke, D., *Philosophical Problems of the Social Sciences.* New York: The Macmillan Company, 1965.

Campbell, D. T. and C. J. Stanley, *Experimental and Quasi-Experimental Designs for Research*. Chicago: Rand McNally & Co., 1967.

Conant, J. B., *Modern Science and Modern Man*. Garden City, N.Y.: Doubleday & Company, Inc., 1953.

Huff, D., *How to Lie with Statistics*. New York: W. W. Norton & Company, Inc., 1954.

Hyman, R., *The Nature of Psychological Inquiry*. Englewood Cliffs, N.J.: Prentice-Hall, Inc., 1964.

Mehlberg, H., *The Reach of Science*. Toronto: University of Toronto Press, 1958.

Snow, C. P., *The Search*. New York: The New American Library, 1934.

Storer, N. W., *The Social System of Science*. New York: Holt, Rinehart & Winston, Inc., 1966.

Taton, R., *Reason and Chance in Scientific Discovery*. New York: Science Editions, Inc., 1962.

Underwood, B. J., *Psychological Research*. New York: Appleton-Century-Crofts, 1957.

White, A. D., *A History of the Warfare of Science with Theology in Christendom*. New York: George Braziller, Inc., 1955.

Index